Principles and Practice of Engineering
ARCHITECTURAL ENGINEERING
SAMPLE QUESTIONS AND SOLUTIONS

SPONSORED BY
Professional Exam Development Subcommittee
Architectural Engineering Institute of ASCE

EDITED BY
Mark McAfee

ASCE American Society
of Civil Engineers
1801 ALEXANDER BELL DRIVE
RESTON, VIRGINIA 20191–4400

Architectural Engineering Institute

Abstract: The Architectural Engineering Institute (AEI) of the American Society of Civil Engineers (ASCE) has prepared this handbook to assist candidates who are preparing for the Principles and Practice of Engineering (PE) examination in architectural engineering. A sample examination prepared by the AEI of ASCE is presented in this book. By illustrating the general content of the subject areas and formats, the questions should be helpful in preparing for the examination. Solutions are presented for all the questions. The solution presented may not be the only way to solve the question. The intent is to demonstrate the typical effort required to solve each question. No representation is made or intended as to future examination questions, content, or subject matter.

Library of Congress Cataloging-in-Publication Data

Principles and practice of engineering : architectural engineering sample questions & solutions / edited by Mark McAfee ; sponsored by the Professional Exam Development Subcommittee, Architectural Engineering Institute of ASCE.
        p. cm.
Includes bibliographical references and index.
  ISBN 0-7844-0657-X
  1. Building--Examinations--Study guides. 2. Building--Problems, exercises, etc. 3. Structural analysis (Engineering)--Examinations--Study guides. 4. Structural analysis (Engineering)--Problems, exercises, etc. I. McAfee, Mark. II. American Society of Civil Engineers.

TH166 .P75 2002
624.1'076--dc21   2002036670

Acknowledgments

The production of this handbook was performed by the Professional Exam Development Subcommittee of the Architectural Engineering Institute (AEI) of the American Society of Civil Engineers (ASCE). Many people assisted and volunteered their services in the development of the architectural engineering principles and practice of engineering examination and this handbook. The professional exam development subcommittee gratefully acknowledges the contributions of all persons involved in this processes and the support of ASCE and the AEI's Board of Governors.

Contents

* Information in sections marked by (*) is used by permission of the National Council of Examiners for Engineering and Surveying (NCEES).

Introduction

One of the functions of the National Council of Examiners for Engineering and Surveying (NCEES) is to develop examinations that are taken by candidates for licensure as professional engineers. The NCEES is was established to assist and support the licensing boards that exist in all states and U.S. territories. The NCEES provides these boards with uniform examinations that are valid measures of minimum competency related to the practice of engineering.

To develop reliable and valid examinations, the NCEES employs procedures using the guidelines established in the *Standards for Educational and Psychological Testing* published by the American Psychological Association. These procedures are intended to maximize the fairness and quality of the examinations. To ensure that the procedures are followed, the NCEES uses experienced testing specialists possessing the necessary expertise to guide the development of examinations using current testing techniques.

The examinations are prepared by committees composed of professional engineers from throughout the nation. These engineers supply the content expertise that is essential in developing examinations. By using the expertise of engineers with different backgrounds such as private consulting, government, industry, and education, the NCEES prepares examinations that are valid measures of minimum competency.

LICENSING REQUIREMENTS

Eligibility
The primary purpose of licensure is to protect the public by evaluating the qualifications of candidates seeking licensure. While examinations offer one means of measuring the competency levels of candidates, most licensing boards also screen candidates based on education and experience requirements. Because these requirements vary among boards, it would be wise to contact the appropriate board. Board addresses and telephone numbers may be obtained by visiting the NCEES Web site at www.ncees.org or calling the NCEES at (800) 250-3196.

Application Procedures and Deadlines
Application procedures for the examination and instructional information are available from individual boards. Requirements and fees vary among the boards, and applicants are responsible for contacting their board office. Sufficient time must be allotted to complete the application process and assemble required data.

DESCRIPTION OF EXAMINATIONS

Examination Schedule
The NCEES PE examination in architectural engineering will be offered in the spring of each year beginning in 2003. Future administration dates are as follows:

Year	Spring Dates
2003	April 11
2004	April 16
2005	April 15
2006	April 21

You should contact your board for specific locations of exam sites.

Examination Content

The 8-hour PE examination in architectural engineering is a no-choice examination consisting of 80 multiple-choice questions. The examination is administered in two 4-hour sessions, each containing 40 questions. Each question has four answer options. The examination specifications presented in this book give details of the subjects covered on the examination.

Typically, all information required to answer a question is provided within the statement of the question itself. The question poses a scenario that an architectural engineer might encounter in practice soon after licensure. In a few instances, two to three questions are grouped within a single scenario with common information appearing before the questions. In these instances, the questions supply any further information specific to the question and define what is expected as a response to the question. In all cases, the correct response requires a calculation and/or conclusion that demonstrates competent engineering judgment.

EXAMINATION DEVELOPMENT

Examination Validity

Testing standards require that the questions on a licensing examination be representative of the important tasks needed for competent practice in the profession. The NCEES establishes the relationship between the examination questions and tasks by conducting an analysis of the profession that identifies the duties performed by the engineer. This information is used to develop an examination content outline that guides the development of job-related questions.

Examination Specifications

The examination content outline presented in this book specifies the subject areas that were identified for architectural engineering and the percentage of questions devoted to each of them. The percentage of questions assigned to each of the subject areas reflects both the frequency and importance experienced in the practice of architectural engineering.

Examination Preparation and Review

Examination development and review workshops are conducted regularly by exam developers. Additionally, workshops are held as required to supplement the bank of questions available. The content and format of the questions are reviewed by the exam developers for compliance with the specifications and to ensure the quality and fairness of the examination. These engineers are selected with the objective that they be representative of the profession in terms of geography, ethnic background, gender, and area of practice.

Minimum Competency

One of the most critical considerations in developing and administering examinations is the establishment of passing scores that reflect a standard of minimum competency. The concept of minimum competency is uppermost in the minds of the developers as they assemble questions for the examination. Minimum competency, as measured by the examination component of the licensing process, is defined as the lowest level of knowledge at which a person can practice professional engineering in such a manner that will safeguard life, health, and property and promote the public welfare.

To accomplish the setting of fair passing scores that reflect the standard of minimum competency, the NCEES conducts passing score studies on a periodic basis. At these studies, a representative panel of engineers familiar with the candidate population uses a criterion-referenced procedure to set the passing score for the examination. Such procedures are widely recognized and accepted for occupational licensing purposes. The panel discusses the concept of minimum competence and develops a written standard of minimum competency that clearly articulates what skills and knowledge are required of licensed engineers. Following this, the panelists take the examination and then evaluate the difficulty level of each question in the context of the standard of minimum competency.

The NCEES does not use a fixed-percentage pass rate such as 70% or 75% because licensure is designed to ensure that practitioners possess enough knowledge to perform professional activities in a manner that protects the public welfare. The key issue is whether an individual candidate is competent to practice and not whether the candidate is better or worse than other candidates.

The passing score can vary from one administration of the examination to another to reflect differences in difficulty levels of the examinations. However, the passing score is always based on the standard of minimum competency. To avoid confusion that might arise from fluctuations in the passing score, scores are converted to a standard scale which adopts 70 as the passing score. This technique of converting to a standard scale is commonly employed by testing specialists.

SCORING PROCEDURES

The examination consists of 80 equally weighted multiple-choice questions. There is no penalty for marking incorrect responses; therefore candidates should answer each question on the examination. Only one response should be marked for each question. No credit is given where two or more responses are marked. The examination is compensatory—poor scores in some subjects can be offset by superior performance elsewhere.

The legal authority for making licensure decisions rests with the individual licensing boards and not with the NCEES. Consequently, each board has the authority to determine the passing score for the examination. The NCEES provides each board with a recommended passing score based on the criterion-referenced procedure described previously.

EXAMINATION PROCEDURES AND INSTRUCTIONS

Examination Materials
Before the morning and afternoon sessions, proctors will distribute examination booklets containing an answer sheet. You should not open the examination booklet until you are instructed to do so by the proctor. Read the instructions and information given on the front and back covers. Listen carefully to all the instructions the proctor reads.

The answer sheets for the multiple-choice questions are machine scored. For proper scoring, the answer spaces should be blackened completely. Since October 2002, the NCEES has provided mechanical pencils with 0.7-mm HB lead to be used in the examination. You are not permitted to use any other writing instrument. If you decide to change an answer, you must erase the first answer completely. Incomplete erasures and stray marks may be read as intended answers. One side of the answer sheet is used to collect identification and biographical data. Proctors will guide you through the process of completing this portion of the answer sheet prior to taking the test. This process will take approximately 15 minutes.

Starting and Completing the Examination
You are not to open the examination booklet until instructed to do so by your proctor. If you complete the examination with more than 30 minutes remaining, you are free to leave after returning all examination materials to the proctor. Within 30 minutes of the end of the examination, you are required to remain until the end to avoid disruption to those still working and to permit orderly collection of all examination materials. Regardless of when you complete the examination, you are responsible for returning the numbered examination booklet assigned to you. Cooperate with the proctors collecting the examination materials. Nobody will be allowed to leave until the proctor has verified that all materials have been collected.

References

The PE examination is open-book. Your board determines the reference materials and calculators that will be allowed. In general, you may use textbooks, handbooks, bound reference materials, and a non-communicating, battery-operated, silent, non-printing calculator. States differ in their rules regarding calculators and references, and you should contact your board for specific advice. AEI has provided a list of references in this book that may be of assistance. The list is intended only for informational purpose and is by no means inclusive or exclusive of references that will assist in test-taking.

Special Accommodations

If you require special accommodations in the test-taking procedure, you should communicate your need to your board office well in advance of the day of the examination so that necessary arrangements may be made.

EXAMINATION SPECIFICATIONS

Examination Specification for the Principles and Practice of Engineering Examination in Architectural Engineering

	Approximate Percentage of Examination
I GENERAL KNOWLEDGE	**12%**

A. Building Systems   5%

1. Functions of Building Systems
   a. Mechanical, HVAC, plumbing, fire protection
   b. Electrical and lighting
   c. Structural
   d. Architectural
2. Fire protection systems relevant to electrical, mechanical, and structural design components.
3. Conditions for retrofit or re-use of existing buildings with respect to system integration.
4. Aspects of building performance that affect human comfort.
5. Knowledge of which building systems or functions are critical in emergencies.
6. Understanding of framing alternatives as they relate to electrical, mechanical, and structural design.

B. Construction and Building Materials   4%

1. Terminology - lighting, electrical, mechanical and structural.
2. Basic construction methods and materials.
3. Knowledge of material behavior and properties (e.g. heat transfer, volume change).

C. Lateral Loads and Displacement Issues   1%

1. Lateral load and displacement effects on mechanical, electrical, and architectural systems.

D. Codes, Regulations, and Statutes   2%

1. Model Codes
2. Code of Ethics

3.

3. ADA and its implications for electrical, mechanical, and structural design.

II CONSTRUCTION MANAGEMENT   **15%**

  A. Economic and Financial Issues   6%
1. Control of Purchase Order, monitoring items relative to lead time and placement of items, etc.
2. Building Cost Estimating
3. Value Engineering
4. Quantity take-off methods
5. Bonding, taxes, profit/overhead and permits on estimate.

  B. Construction Processes   4%
1. Erection Processes and Sequence
2. Administration of Information (e.g.: differing site conditions, change orders, etc.).

  C. Project Management   5%
1. Scheduling (Sequence of Activities)
2. Quality Control
3. Alternative Project Delivery Systems

III ELECTRICAL AND LIGHTING SYSTEMS   **23%**

  A. Theory   9%
1. Electrical power circuit theory
2. Short circuit theory
3. Load flow
4. Power Factor Correction
5. Overcurrent protection device coordination

  B. Basic Electrical Knowledge   14%
1. Grounding systems
2. Basic electrical construction methods and materials
3. Overcurrent protection
4. Branch circuit and feeder conductor sizing
5. Lighting system design

6. HVAC system power distribution issues
7. Emergency systems i.e. exit lighting, batteries, generators, alarm systems, etc.
8. Codes
   a. National Electric Code
   b. Lighting Energy Code - ASHRAE 90.1

IV   MECHANICAL SYSTEMS   23%

A. Theory   9%
   1. Fan Laws
   2. Pump Laws
   3. Psychometrics
   4. Air Pressure Drop (Static Pressure/Velocity Pressure)
   5. Humidity/Latent Load

B. Basic Mechanical Knowledge   14%
   1. Types of Pumps and Pumping Systems
   2. Fire Protection Sprinkler and Standpipe Classification
   3. Steam Pressure Classification
   4. Pipe Expansion
   5. Duct Materials
   6. Insulation
   7. Valves and Fittings
   8. Outside Air Quantities and Ventilation
   9. Combustion Air Requirements and Sizing
   10. HVAC System Types
   11. Chiller Types and Applications
   12. Boiler Types and Applications
   13. Criteria for Diffuser/Register/Grille Selection
   14. Sanitary Waste and Vent Systems
   15. Domestic Water
   16. Roof Drainage

V   STRUCTURAL SYSTEMS   27%

A. Loads and Analysis   11%

1. Types and methods of calculating design loads (e.g. wind, live).
2. Different types of analysis and design philosophies or methods.
3. Material and member behavior in response to specific loading conditions.
4. Structural serviceability issue requirements (e.g. vibrations and deflections).

B. Design   16%

1. Design and construction requirements for fire (e.g. firewalls, material fire rating).
2. Material specific code requirements.
   a. Steel
   b. Concrete
   c. Timber
   d. Masonry
3. Structural Systems
4. Elastic vs. Plastic Design Procedures
5. Connections
   a. Timber
   b. Steel
   c. Concrete
6. Foundation Systems

Total =   100%

SAMPLE QUESTIONS FOR THE MORNING PORTION
OF THE EXAMINATION
IN ARCHITECTURAL ENGINEERING

1. In a high rise office building, which of the following groups of systems are typically required by building codes to have emergency power for life safety reasons?

    a. Elevator shaft exhaust fans, egress corridor lighting, temperature controls.
    b. Stair pressurization fans, exit lights, egress corridor lighting.
    c. Fire alarm system, egress corridor lighting, telephone system.
    d. Fire alarm system, exit lights, exit stair exhaust fans.

2. When selecting the roof location for a plumbing vent, which of the following design elements should be taken into consideration?

    a. The location of exhaust fan outlets.
    b. The location of other plumbing vents.
    c. The location of roof-top fire hose connections.
    d. The location of outside air intakes for the HVAC system.

3. Which of the following DOES NOT influence the structural performance of a reinforced concrete beam during a fire?

    a. Concrete coverage of reinforcing.
    b. Density of concrete.
    c. Concrete compressive strength.
    d. Concrete aggregate type.

4. In general, which of the following is the LEAST efficient frame in limiting drift from lateral loads?

    a. Moment resisting frame.
    b. X-Braced Frame.
    c. K-Braced Frame
    d. Shear Wall

5. In a high rise building, the HVAC system should respond to a fire alarm in which of the following ways:

    a. Pressurize the elevator shafts and exit stairways.
    b. Pressurize only the elevator shafts.
    c. Exhaust the elevator shafts.
    d. Exhaust the exit stairways.

For **Question 6**, refer to the following information.

Previous project information:

Location	Year Constructed	Number of Cars	Cost
Charlotte, NC	1998	580	$2,726,000
Birmingham, AL	1996	560	$3,416,000
Jacksonville, FL	1995	630	$3,213,000

Geographic Cost Indexes

Atlanta, GA	90.3
Birmingham, AL	88.4
Charlotte, NC	79.1
Jacksonville, FL	84.4

Historical Cost Indexes

1994	87.1
1995	89.3
1996	90.8
1997	95.1
1998	97.4
1999	100.0

Assume inflation after 1999 is 3.5% per year.

6. Using the given historical cost information, the projected Year 2003 cost for a 600 car above grade parking garage in Atlanta, Georgia is most nearly?

   a. $3,180,000
   b. $3,688,300
   c. $3,825,500
   d. $4,243,200

7. In a construction contract utilizing the 1987 or 1997 Edition AIA/A201 form of general conditions, a Construction Change Directive (CCD) has been issued to the Contractor. Issuance of the CCD directs the Contractor to proceed with the work in which of the following ways:

   a. Proceed on a time and material basis.
   b. Proceed and establish the final price of the change.
   c. Proceed and establish a time and a method to finalize the price.
   d. Proceed once a price has been finalized.

For **Question 8**, refer to the following information.

<u>Material and Labor Cost Information:</u>

Concrete masonry units	$1.85 each
Horizontal joint reinforcing	$0.18/lineal foot
Mortar	$0.35/sf of wall
Labor crew cost	$265/hr
Crew productivity	310 sf/day
Work Day	8 Hours

8. Using the given material and labor information, the cost to construct 800 lineal feet of a 12 foot tall concrete masonry unit wall, using standard face size 12 inch thick concrete masonry units with horizontal joint reinforcing at every course is most nearly:

    a. $83,950
    b. $88,370
    c. $89,230
    d. $91,440

9. A construction contract is issued to a general contractor under an 1987 or 1997 Edition AIA/A201 form of contract. During the course of construction the architect believes that some of the work installed is in non-compliance with the technical specifications of the project. The architect directs the contractor to remove a portion of the work to expose the work in question, to determine if the hidden work was in compliance with the specifications. The work exposed is found to be in non-compliance. The architect has the financial responsibility for:

    a. the reinstallation cost of the work covering the non-complying work.
    b. the removal cost of the work covering the non-complying work.
    c. any testing and inspection cost to find if the work was in non-compliance.
    d. none of the costs associated with the removal, testing, or reinstallation.

10. Based on the information provided and assuming an annual interest rate of 5%, the roofing system which provides the lowest cost alternative is:

   a. EPDM roofing - Initial cost is $50,000. Expected maintenance costs are $1,200 per year. Expected life is 15 years.
   b. Modified bitumen roofing - Initial cost is $70,000. Expected maintenance costs are $600 per year. Expected life is 20 years.
   c. Built-up roofing - Initial cost is $80,000. Expected maintenance costs are $600 per year. Expected life is 25 years.
   d. Metal roofing - Initial cost is $120,000. Expected maintenance costs are $200 per year. Expected life is 40 years.

For **Question 11**, refer to the diagram below.

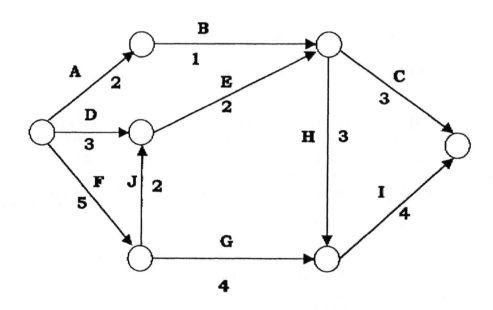

11. Using the given schedule diagram, the total float in the path denoted by activities A — B — C is:

   a. 4
   b. 6
   c. 7
   d. 10

12.

12. In a balanced 3 phase circuit, the voltage, V_{LN}, equals 120V $\angle 0°$. The phase sequence is "a","b","c". The current (A) in I_{LN} of a wye-connected load having a $Z_L = 2\Omega\angle 40°$ (ohms) is most nearly:

a. 60$\angle$ 40°
b. 60$\angle$-40°
c. 240$\angle$40°
d. 240$\angle$-40°

For **Question 13**, refer to the table below.

MOTOR FULL LOAD CURRENTS
(Amperes)

HP	1 Φ				3 Φ				
	115 V	120 V	208 V	230 V	208 V	230 V	460 V	480 V	575 V
1/33	1.7	1.7	1.0	0.9	--	--	--	--	--
1/20	1.9	1.9	1.1	1.0	--	--	--	--	--
1/15	2.1	2.0	1.2	1.0	--	--	--	--	--
1/12	2.3	2.2	1.3	1.2	--	--	--	--	--
1/8	3.1	3.0	1.7	1.6	--	--	--	--	--
1/6	4.4	4.2	2.4	2.2	--	--	--	--	--
1/4	5.8	5.6	3.2	2.9	--	--	--	--	--
1/3	7.2	6.9	4.0	3.6	--	--	--	--	--
1/2	9.8	9.4	5.4	4.9	2.4	2.2	1.1	1.1	0.9
3/4	--	--	7.6	6.9	3.5	3.2	1.6	1.5	1.3
1	--	--	8.8	8.0	4.6	4.2	2.1	2.0	1.7
1-1/2	--	--	11.0	10.0	6.6	6.0	3.0	2.9	2.4
2	--	--	13.2	12.0	7.5	6.8	3.4	3.3	2.7
3	--	--	18.7	17.0	10.6	9.6	4.8	4.6	3.8
5	--	--	30.8	28.0	16.8	15.2	7.6	7.3	6.1
7-1/2	--	--	44.0	40.0	24	22	11	11	9
10	--	--	55.0	50.0	31	28	14	13	11
15	--	--	--	--	46	42	21	20	17
20	--	--	--	--	60	54	27	26	22
25	--	--	--	--	75	68	34	33	27
30	--	--	--	--	88	80	40	38	32
40	--	--	--	--	115	104	52	50	42
50	--	--	--	--	144	130	65	62	52

13. A single motor branch circuit is to be used to serve three motors rated at 1/3HP, 1/2HP, and 1/4HP. The motors are operating on a 115 volt, 60 hertz, single phase circuit. Using the given table, the calculated full load amps (A) that will be used to size the feeder for this branch circuit according to the National Electric Code is most nearly:

   a. 22.8
   b. 25.3
   c. 27.7
   d. 29.4

14.

For **Question 14**, refer to the following information.

Given Information:

	Phase A	Phase B	Phase C
Lighting	50 kVA	40 kVA	45 kVA
Receptacles	10 kVA	20 kVA	15 kVA
Other Continuous Loads	20 kVA	25 kVA	25 kVA

14. A three phase panel for an office building is described above. Using the National Electric Code defined demand factors, the total kVA load on this panel is most nearly:

a. 232.5
b. 250.0
c. 273.75
d. 301.25

For **Question 15**, refer to the following information.

Given Information:

3 phase
13,200-277/480V
1500 kVA
5.5% Impedance

15. For the transformer described above, the full load current (A) available at the secondary terminals of the transformer is most nearly:

a. 27,272
b. 5,415
c. 3,125
d. 1,804

For **Question 16**, refer to the photometric data sheet below.

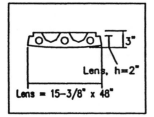

PHOTOMETRIC DATA
Luminaire — "BAE3T8A"

DESCRIPTION:
__15 x 48 Prismatic Fluo.__
__Wraparound with__
__Metal Ends__
LAMP: __3F32T8/35__ VOLTS __120__
TF: __1.0__ BF: __.88__ Watts: __87__
LUMEN/LAMP __2900__ DATE __2000__
SHIELDING ANGLE: N __90__ L __90__
BARE __2350__ FILE __BAE332__
TEST NO. __8365TL__ BY: __RNH__

LUMINOUS INTENSITY

	ALONG	22.5	45.0	67.5	ACROSS
0	2237	2237	2237	2237	2237
5	2229	2235	2239	2233	2224
15	2150	2166	2184	2197	2192
25	1987	2019	2073	2112	2112
35	1706	1764	1856	1898	1895
45	1295	1342	1421	1478	1486
55	822	832	871	912	922
65	467	421	430	496	568
75	261	244	255	365	442
85	83	112	202	340	391
90	7	52	176	325	371
95	5	41	164	291	329
105	19	54	157	233	260
115	27	52	121	188	199
125	38	59	95	125	142
135	48	59	66	91	97
145	46	57	61	66	61
155	40	57	55	59	61
165	38	34	43	43	40
175	30	33	30	32	35
180	32	32	32	32	32

OUTPUT DATA

ZONES	LUMENS	%TOT. LAMP	%TOT. DIST.
0– 40°	2935		
40– 90°	2900		
90–180°	669		
0–180°	6504		

LUMINANCE DATA

ANGLE	ALONG MAX.	ALONG AVE.	ALONG M/A	ACROSS MAX.	ACROSS AVE.	ACROSS M/A
45	2670			3670		
55	2052			2540		
65	1455			1825		
75	1045			1650		
85	785			1535		

ZONES – 5 degrees

DOWN HEMISPHERE	UP HEMISPHERE	ZONAL CONSTANT
0 – 5	175 –	0.024
5 – 10	170 – 175	0.072
10 – 15	165 – 170	0.119
15 – 20	160 – 165	0.165
20 – 25	155 – 160	0.210
25 – 30	150 – 155	0.253
30 – 35	145 – 150	0.295
35 – 40	140 – 145	0.334
40 – 45	135 – 140	0.370
45 – 50	130 – 135	0.404
50 – 55	125 – 130	0.435
55 – 60	120 – 125	0.462
60 – 65	115 – 120	0.486
65 – 70	110 – 115	0.506
70 – 75	105 – 110	0.523
75 – 80	100 – 105	0.535
80 – 85	95 – 100	0.543
85 – 90	90 – 95	0.548

ZONAL CONSTANTS

ZONES – 10 degrees

DOWN HEMISPHERE	UP HEMISPHERE	ZONAL CONSTANT
0 – 10	170 – 180	0.095
10 – 20	160 – 170	0.283
20 – 30	150 – 160	0.463
30 – 40	140 – 150	0.628
40 – 50	130 – 140	0.774
50 – 60	120 – 130	0.897
60 – 70	110 – 120	0.993
70 – 80	100 – 110	1.058
80 – 90	90 – 100	1.091

16. Using the given photometric data sheet, the value of the "Zonal Lumens" for the elevation angle of 75 degrees from nadir is most nearly:

    a. 303.9
    b. 321.5
    c. 331.6
    d. 350.8

For **Question 17**, the following balanced 3 phase loads are connected to a 3 phase, 4 wire, 480V/277, 60 Hertz distribution:

    1. 4-three phase induction motors of 10 HP each operating at full load with 90% efficiency and 85% PF.
    2. 1-8 kVA heating load at a Lagging, PF = .92.
    3. 3-single phase, 10 kW lighting loads each at unity power factor.

17. The amount of Capacitive Reactance power, in kVAR, that must be added to the system to yield an overall system power factor of 98% Lagging is most nearly:

    a. 16.65
    b. 21.5
    c. 26.5
    d. 31.35

For **Question 18**, refer to the phasor diagrams below.

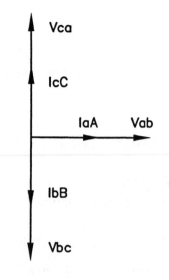

Vector Diagram A

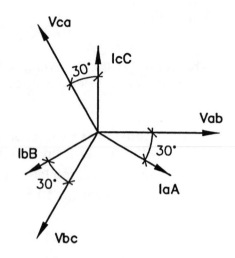

Vector Diagram B

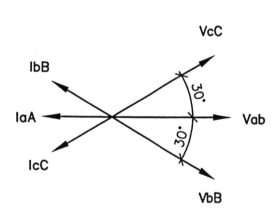

Vector Diagram C

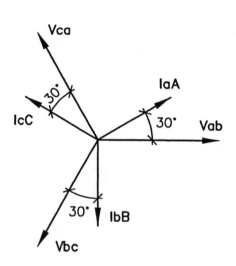

Vector Diagram D

18.

18. Given the above phasor diagrams for a 3 phase system, where the current lags the voltage by 30°, the correct phasor diagram is most nearly:

a. Vector diagram A
b. Vector diagram B
c. Vector diagram C
d. Vector diagram D

19. For rigid metal conduit, according to the National Electric Code, the correct statements regarding bending requirements are most nearly:

I. There shall be not more than the equivalent of four quarter bends between pull points.
II. There shall be not more than 180 degrees of total bends between pull points.
III. There shall be not more than four bends of any angle between pull points.
IV. There shall be not more than 360 degrees in bends between pull points.
V. There shall be not more than the equivalent of four 45 degree bends between pull points.

a. I and IV only
b. I,III and IV only
c. II,III and V only
d. II and V only

For **Question 20**, refer to the Photometric Data Sheet and the design factors given below.

PHOTOMETRIC   DATA

Luminaire —"BAE3T8B"

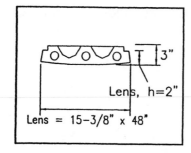

3"

Lens, h=2"

Lens = 15-3/8" x 48"

DESCRIPTION:
  15 x 48 Prismatic Fluo.
  Wraparound with
  Metal Ends
LAMP: 3F32T8/35 VOLTS 120
TF: 1.0 BF: .88 Watts: 87
LUMEN/LAMP 2900 DATE 2000
SHIELDING ANGLE:N 90 L 90
BARE 2350 FILE BAE332
TEST NO. 8365TL BY: RNH

CATEGORY V

OUTPUT DATA

ZONES	LUMENS	%TOT. LAMP	%TOT. DIST.
0— 40°	2935		
40— 90°	2900		
90—180°	669		
0—180°	6504		

LUMINANCE DATA

ANGLE	ALONG			ACROSS		
	MAX.	AVE.	M/A	MAX.	AVE.	M/A
45	2670			3670		
55	2052			2540		
65	1455			1825		
75	1045			1650		
85	785			1535		

FLOOR, ρ_{fc}	COEFFICIENT OF UTILIZATION, ρ_{fc}= 20%									
CLNG., ρ_{cc}	80			70			50			0
WALL, ρ_w	70	50	30	70	50	30	50	30	10	0
ROOM CAVITY RATIO — 1	80	77	74	77	74	71	70	67	65	57
2	74	68	63	71	68	62	62	59	56	49
3	68	61	55	65	59	54	56	51	48	43
4	63	54	48	60	53	47	50	45	42	37
5	57	49	42	55	47	41	45	40	36	32
6	53	44	37	51	43	37	41	35	32	28
7	49	39	33	47	38	33	37	32	28	25
8	45	35	29	44	35	29	33	28	24	22
9	42	32	26	40	31	25	30	25	21	18
10	39	29	23	37	28	23	27	22	19	16

Luminaire Spacing Criterion = 1.2/1.3

20.

Given Information:

Room: Width = 14 Feet   Length =28 Feet   Floor-to-ceiling = 8 Feet

Workplane = 2.5 Feet

Surface Mounted Luminaire, hcc = 0

Surface Reflectances:
  Ceiling = 70%   Walls = 50%   Effective Reflectance of the Floor Cavity,
                                      ρ_{fc} = 20%

Recommended Illuminance Level, E_r = 75 fc

Maintenance Data: LLD = .86   LDD = .70

20. Using the given Photometric Data sheet and design factors, the number (round to whole number) of luminaires required to produce the average, uniform, maintained illuminance level is most nearly:

    a.   11
    b.   12
    c.   13
    d.   14

21. Which of the following is NOT a scheme used to protect cooling towers from freezing?

    a.   Variable frequency drive on the condenser pump
    b.   Glycol in the condenser water
    c.   Condenser by-pass with a remote sump
    d.   Heat trace the sump with basin heaters

22. A building that is to have a room temperature of 75 degrees F, has a supply air temperature of 105 degrees F. If the net heat loss from the building is 162,000 BTU per hour, the air flow (CFM) that must be supplied is most nearly:

    a.   5,400
    b.   5,000
    c.   2,160
    d.   1,542

For **Questions 23-24** , refer to the following information.

A mechanically induced draft cooling tower is used to reject heat from an air conditioning system's water chiller. The chiller has a capacity of 600 tons of refrigeration.

23. At 75% load, the energy input is 0.70 kW per ton. The heat rejected to the cooling tower (Btu/hr) at this condition is most nearly:

    a. 5,056,000
    b. 5,400,000
    c. 6,475,000
    d. 7,550,000

24. It is desired to use a two-speed motor on the cooling tower fan to save energy at part load conditions. The best way to save energy is achieved by controlling the fan speed from:

    a. Cooling tower entering air temperature
    b. Cooling tower leaving water temperature
    c. Leaving chilled water temperature
    d. Condenser leaving water temperature

For **Question 25**, refer to the roof plan, riser diagram, and tables below.

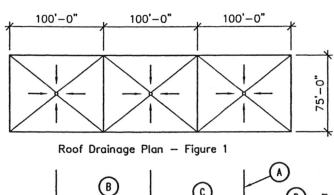

Roof Drainage Plan — Figure 1

Storm Riser Diagram — Figure 2

The roof plan for a building with three internal storm drains is shown in Figure 1, while the storm riser diagram in shown in Figure 2. The four pipe sections are labeled in the riser diagram, i.e. A, B, C, and D. The pipe section B, C, D is sloped downward at a rate of ¼ inches per foot.

From Dagostino: "Mechanical and Electrical Systems in Construction and Architecture," 1995. Reprinted by permission of Pearson Education, Inc., Upper Saddle River, NJ.

Size of Pipe (inches)	Maximum Rainfall in Inches per Hour				
¼" /ft Slope	2	3	4	5	6
3	2320	1546	1160	928	773
4	5300	3533	2650	2120	1766
5	9440	6293	4720	3776	3146
6	15100	10066	7550	6040	5033

Table 1: Size of Horizontal Rainwater Piping
(Quantities are Projected Roof Area in Square Feet)

Rainfall (in./hour)	Size of Drain or Leader in Inches				
	2	3	4	5	6
2	1440	4400	9200	17300	27000
3	960	2930	6130	11530	17995
4	720	2200	4600	8650	13500
5	575	1760	3680	6920	10800
6	480	1470	3070	5765	9000

Table 2: Size of Vertical Conductors or Leaders
(Quantities are Projected Roof Area in Square Feet)

25. Using the given roof plan, riser diagram, and tables, and assuming a rainfall rate of 5 inches per hour, the required nominal size (inches) of the vertical riser, Section A, is most nearly:

    a. 2
    b. 3
    c. 4
    d. 6

For **Question 26**, refer to the section below.

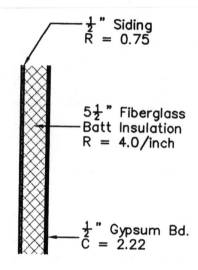

½" Siding
R = 0.75

5½" Fiberglass
Batt Insulation
R = 4.0/inch

½" Gypsum Bd.
C = 2.22

26. In the given wall section, assume that there are 2" by 6" wood studs at 24" on center (R= 1.0 / inch). Neglect the interior and exterior film coefficients. The overall U-value for the given wall-section is most nearly:

    a. 0.033
    b. 0.043
    c. 0.045
    d. 0.149

27. Ordinary-temperature sprinklers are rated for use in what temperature range?

    a. 135°F to 170°F
    b. 175°F to 225°F
    c. 250°F to 300°F
    d. 350°F to 450°F

24.

For **Question 28**, refer to the sketch and Friction Loss Chart below.

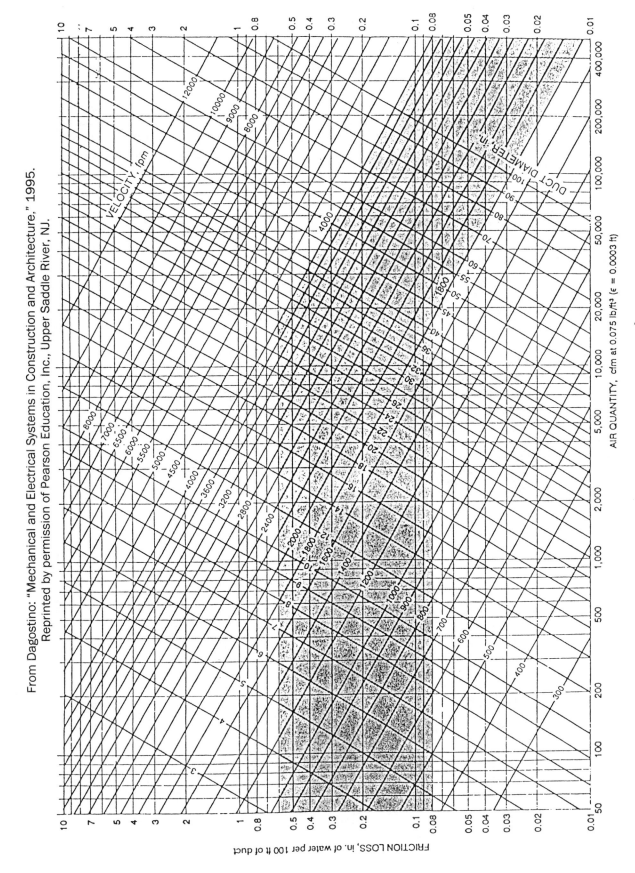

Fig. 9-2 Friction Chart for Round Duct ($\rho = 0.075\ lb_m/ft^3$ and $e = 0.0003\ ft$)

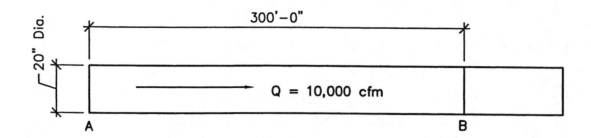

300'-0"

20" Dia.

Q = 10,000 cfm

A

B

28. Using the given sketch and the Friction Loss Chart, if the total pressure at point A is 6.00 inches of water gauge, the static pressure in the round duct at Point B is most nearly:

   a. 1.14 inches of water gauge
   b. 2.4 inches of water gauge
   c. 3.5 inches of water gauge
   d. 4.8 inches of water gauge

29. The best way to control the fan supply pressure for a variable air volume HVAC ` System is to:

   a. Install a static pressure sensor 2/3 of the way down the main duct.
   b. Install a static pressure sensor at the discharge of the supply fan.
   c. Install a static pressure sensor at the fan inlet and discharge.
   d. Install a static pressure sensor at the fan inlet.

For **Question 30**, refer to the roof plan below.

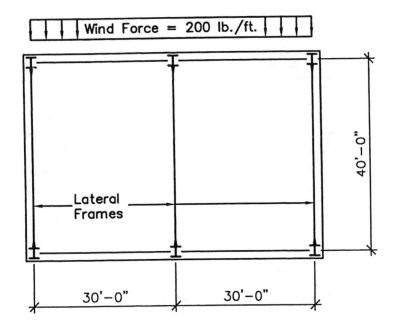

30. Using the given roof plan and a lateral wind force of 200#/ft. applied at the roof level, the maximum required capacity of the flexible roof diaphragm (#/ft.) is most nearly:

 a. 50
 b. 75
 c. 100
 d. 150

31. A 7 feet by 7 feet square footing resists a service axial load of 100k and a service overturning moment of 50 ft-k. The maximum soil pressure (ksf) under the footing, neglecting the weight of the footing, is most nearly:

 a. 1.17
 b. 2.04
 c. 2.92
 d. 3.06

For **Question 32**, refer to the sketch below.

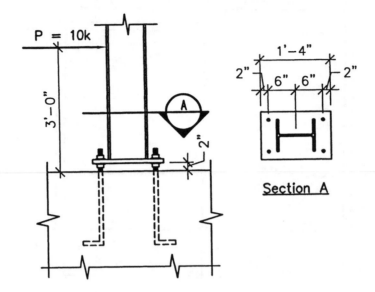

P = 10k

3'-0"

A

2"

1'-4"

2"   6"   6"   2"

Section A

32. The base plate grout was left out during construction. The tension (kips) in a <u>single</u> anchor bolt at the bottom of the base plate due to the overturning force "P" as shown on the given sketch is most nearly:

a. 12.1
b. 14.2
c. 15.0
d. 28.3

For **Question 33**, refer to the diagrams below.

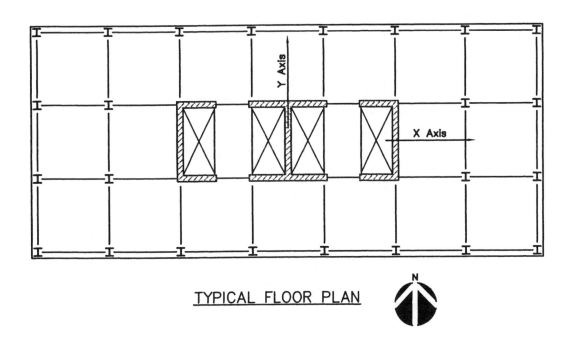

TYPICAL FLOOR PLAN

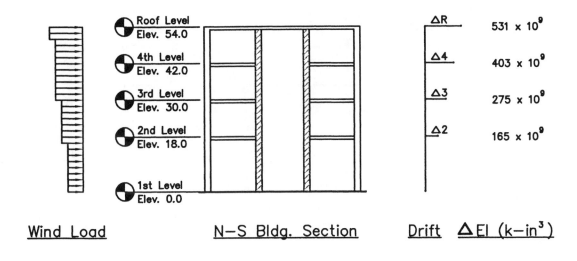

Wind Load   N—S Bldg. Section   Drift $\triangle$ EI (k—in^3)

33. Assume the normal weight concrete shear walls as shown on the given diagrams provides all of the stiffness for drift checks and their combined moment of inertia about the x-axis is 2,190 ft^4. If all floors and walls have the same concrete compressive strength, the <u>MINIMUM</u> f'c (psi) needed to limit the inter-story drift to 0.005h is most nearly:

    a.   3,000
    b.   4,000
    c.   5,000
    d.   6,000

For **Question 34**, refer to the diagram below.

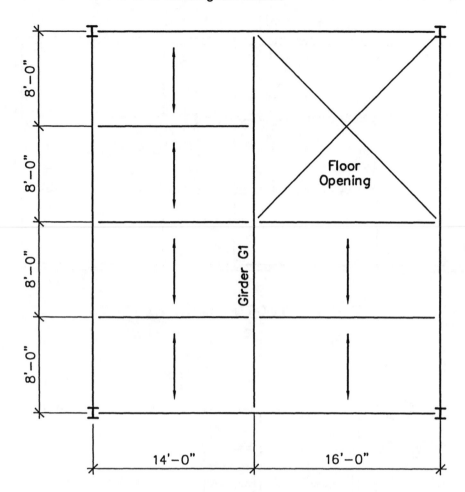

Indicates Direction of
One—Way Conc. Slab on
Metal Form Deck

$R = 0.08 \, (\, A{-}150 \,)$ Where

R = Reduction in Percentage

A = Area of Floor Supported by Girder G1 (Sq. Ft.)

34. Using the given floor plan and live load reduction formula, the live load reduction factor R (%) for girder G1 is most nearly:

   a. 7
   b. 9
   c. 11
   d. 16

For **Questions 35-37**, refer to Figure 1.

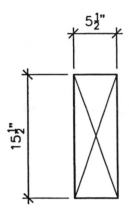

$5\frac{1}{2}$"

$15\frac{1}{2}$"

Wood Beam Section

Given Information:

Wood beam is oriented such that loads are applied in the direction of the beams strong axis

Wood beam is in conditioned environment

Visually graded lumber

Wood beam is seasoned lumber

F_b = 1,300 psi

F_v = 140 psi

E = 1,500,000 psi

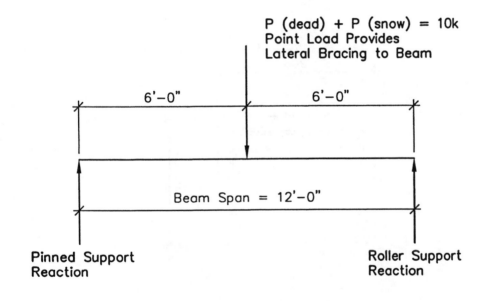

P (dead) + P (snow) = 10k
Point Load Provides
Lateral Bracing to Beam

6'-0"   6'-0"

Beam Span = 12'-0"

Pinned Support
Reaction

Roller Support
Reaction

FIGURE 1

35. Using the information given in Figure 1, and the given beam stability factor, C_L, of 0.98, the allowable bending stress (psi) is most nearly:

   a.   1,274
   b.   1,326
   c.   1,421
   d.   1,465

36. Using the information given in Figure 1, the slenderness ratio for the given wood beam is most nearly:

    a. 6.4
    b. 8.2
    c. 8.6
    d. 9.1

37. Using the information given in Figure 1, if the shear to be resisted by the beam shown is 5 kips, the ratio of actual shear stress to the allowable shear stress is most nearly:

    a. .37
    b. .42
    c. .55
    d. .63

For **Question 38**, refer to the section below.

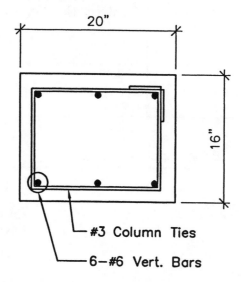

Conc. Column Section

38. Using the given concrete column section, the minimum required tie spacing (inches) is most nearly:

    a. 18
    b. 16
    c. 12
    d. 8

32.

For **Questions 39-40**, refer to the truss geometry and load conditions below.

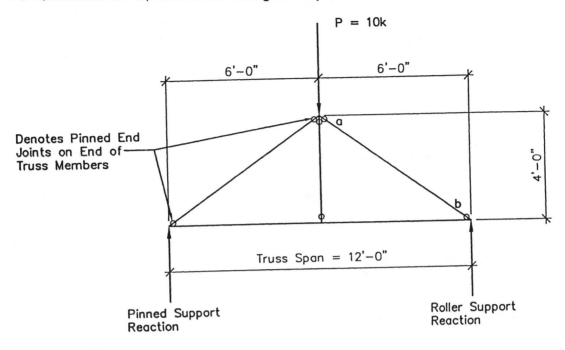

Steel Truss Diagram — Figure 1

39. Using the given truss geometry and load conditions in Figure 1, the force (kips) in member a—b of the truss is most nearly:

    a. 5.0
    b. 6.0
    c. 7.5
    d. 9.0

40. Using the given truss geometry in Figure 1 and the following given information, if the force in member a—b is 25k compression with both ends having a pinned connection and assuming the end connection detail does not reduce the capacity of the member, the lightest weight double angle capable of resisting the member force is most nearly:

    Given Information:

    Fy=36 ksi.   Use ASD methods.

    a. 2-L2x2x3/8
    b. 2-L2½x2½x5/16
    c. 2-L3x3x3/16
    d. 2-L3x3x5/16

33.

SOLUTIONS FOR SAMPLE QUESTIONS
FOR THE MORNING PORTION
OF THE EXAMINATION
IN ARCHITECTURAL ENGINEERING

SOLUTIONS FOR SAMPLE QUESTIONS FOR THE MORNING PORTION OF THE EXAMINATION IN ARCHITECTURAL ENGINEERING

1. Stair pressurization fans, exit lights, egress corridor lighting.

    Correct Answer is: b

2. The location of outside air intakes for the HVAC system.

    Correct Answer is: d

3. Concrete compressive strength

    Correct Answer is: c

4. Moment resisting frame.

    Correct Answer is: a

5. Pressurize the elevator shafts and exit stairways.

    Correct Answer is: a

6. Index for 2003 = $100.00 \times 1.035^4 = 114.8$

    2003 Atlanta Cost per Car

    Charlotte $2{,}726{,}000/580 = \$4{,}700/car \times \dfrac{114.8}{97.4} \times \dfrac{90.3}{79.1} = \$6{,}324/car$

    Birmingham $3{,}416{,}000/560 = \$6{,}100/car \times \dfrac{114.8}{90.8} \times \dfrac{90.3}{88.4} = \$7{,}878/car$

    Jacksonville $3{,}213{,}000/630 = \$5{,}100/car \times \dfrac{114.8}{89.3} \times \dfrac{90.3}{84.4} = \$7{,}015/car$

    Average cost of three projects $\dfrac{(6324 + 7878 + 7015)}{3} = \$7{,}072.00/car$

    Cost of project = 7,072 x 600 = $4,243,200.00

    Correct Answer is: d

7. The CCD obligates the contractor to proceed with the work, and establishes a method and time to settle the final price. This way the settlement of the final price causes no negative impact on the project schedule.

    Correct Answer is: c

8. Area of Wall: 800 x 12 = 9,600 Square Feet

    Horizontal Joint Reinforcing: 12 feet tall wall = 18 block courses = 17 rows of joint reinforcing   17 x 800 =   13,600 lineal feet

    Cost of CMU/SF = $1.85/block x $\dfrac{12 \times 12}{8 \times 16}$ = $2.08 Square Foot

    Material Costs

Concrete Masonry Units	9,600 Square Feet x 2.08	=	19,968.00
Mortar	9,600 Square Feet x 0.35	=	3,360.00
Horizontal Joint Reinforcing	13,600 Lineal Feet x 0.18	=	2,448.00
			$25,776.00

    Labor Costs

    Cost per SF = $\dfrac{265}{Hour}$ x $\dfrac{8.0\ Hours}{Day}$ x $\dfrac{1\ Day}{310\ Square\ Feet}$ = $6.84 Square Feet

    9,600 x 6.84 = $65,664.00

    Total Cost = 25,776.00 + 65,664.00 = $91,440.00

    Correct Answer is: d

9. None of the cost associated with the removal, testing, or reinstallation.

    Correct Answer is: d

10. Equivalent Uniform Annual Cost = Annual Capital Return + Annual Fixed Costs

The formula for capital recovery is: $\dfrac{i(1+i)^n}{(1+i)^n-1}$

i = *interest rate*
n = *number of years*

a.	EPDM Roofing	:	50,000 x 0.0963 + 1200	=	$6,015.00
b.	Modified Bitumen Roofing	:	70,000 x 0.0802 + 600	=	$6,214.00
c.	Built-up Roofing	:	80,000 x 0.071 + 600	=	$6,280.00
d.	Metal Roofing	:	120,000 x 0.0583 + 200	=	$7,196.00

Correct Answer is: a. EPDM Roofing

11. The critical path for the schedule diagram is 16 time units. The path is defined as F - J - E - H -I. The path A - B - C has a duration of 6 time units. The difference between the critical path and the defined path is 16 - 6 = 10 time units.

Correct Answer is: d

12. $V = IZ$ $I_{LN} = \dfrac{V_{LN}}{Z_L} = \dfrac{120\angle 0°}{2\angle 40°} = 60\angle\text{-}40°$

Correct Answer is: b

13. (1.25 x 9.8) + (7.2) + (5.8) = 25.3 A

Correct Answer is: b

14.
Lighting	:	50 + 40 + 45 = 135 kVA x 100%	=	135.0 kVA
Receptacles	:	10 + 20 + 15 = 45 kVA, (45-10) x .50 + 10	=	27.5 kVA
Other Cont. Loads:	:	20 + 25 + 25 = 70 kVA x 100%	=	70.0 kVA
				232.5 kVA

Correct Answer is: a

15. FLA $1500 / \left(\sqrt{3 \times 0.48}\right) = 1{,}804$ =   kVA (3 phase) /SQRT (3) x KV (L-Lsec)

FLA = 1804A

Correct Answer is: d

39.

16.   I average = $\dfrac{261 + (2 \times 244) + (2 \times 255) + (2 \times 365) + 442}{8}$ = 303.9

Zonal Constant = 2 x 3.14159 (cos 70 - cos 80) = 1.058
Zonal Lumens   = I average x zonal constant = 303.9 x 1.058 = 321.5

   Correct Answer is: b

17. $\dfrac{746HP}{W}$ x 10 HP x .90 = $\dfrac{6714W}{.85}$ = 7899 VA

4 x 7899 = 31,596 VA

Theta = cos - 1 (.85) = 31.79

   1.   kW = 4 x 6.714 = 26.86 kW
        kVAR = 31.596 x sin 31.79 = 16.65 kVAR (Lag)
        Theta = cos -1 (.92) = 23.07°
   2.   kW = 80 x cos 23.07 = 73.60 kW
        kVAR = 80 x sin 23.07 = 31.35 kVAR (Lag)
   3.   kW = 3x10 = 30 kW
        kVAR = 0

Sum kW = 26.86 + 73.60 + 30 = 130.45 kW
Theta = cos - 1 (.98) = 11.48°
kVAR @ .92 = 130.46 kw x tan 11.48 = 26.5 kVAR
CAP kVAR = [(16.65 + 31.35) - 26.5] = 21.5 (Leading)

   Correct Answer is: b

18.   Vector Diagram B

   Correct Answer is: b

19.   For rigid metal conduit, there shall be not more than the equivalent of four quarter bends between pull points and there shall be not more than 360 degree in bends between pull points.

   Correct Answer is: a, I and IV only

20. $\rho cc = 70\%$ $\rho w = 50\%$ $\rho fc = 20\%$ $hrc = 8 - 2.5 = 5.5$ feet

$$RCR = \frac{5 \times 5.5 \times (14 + 28)}{14 \times 28} = 2.95 = 3 \qquad CU = .59$$

$$\text{Number of Luminaries} = \frac{75 \times (14 \times 28)}{3 \times 2900 \times .59 \times .70 \times .86 \times .88 \times 1.0} = 10.8$$

11 Luminaries required

Correct Answer is: a

21. Variable frequency drive on the condenser pump.

Correct Answer is: a

22. Using the sensible heat equation

162,000 BTU/hr. = 1.08 x CFM x (30)
 CFM = 5,000

Correct Answer is: b

23. Q condenser = (0.70) (.75 x 600 tons) (3,413 BTU/KW - hr) + (.75 x 600 tons)
(12,000 BTU/hr - ton) = 6,475,095 BTU/hr

Correct Answer is: c

24. The best way to save energy on the cooling tower fan at part load is achieved by controlling fan speed from cooling tower leaving water temperature.

Correct Answer is: b

25. The area of roof which drains to the vertical riser Section A = 100 x 75 = 7,500 Square Feet. Using Table 2, for a 5 inch/hour rainfall with 7,500 Square Feet roof area per vertical riser, a 6 inch vertical riser is required.

A 6 inch vertical riser can accommodate 10,800 Square Feet.

Correct Answer is: d

26. Calculate R-value between studs

1/2 inch Siding			.75
Batt Insulation	5.5 x 4.0	=	22.0
1/2 inch Gyp Board	1/2.22	=	.45
			23.20

Calculate R-value at studs

1/2 inch Siding			.75
Wood Stud	5.5 x 1.0	=	5.5
1/2 inch Gyp Board	1/2.22	=	.45
			6.70

1½ inch wide studs at 24 inches on center.

Wall with Batt Insulation   =   22.5/24   =   .9375

Wall with Wood Stud   =   1.5/24   =   .0625

R = .9375 x 23.20 + .0625 x 6.70 = 22.17

U - value = 1/R   1/22.17 = .0451   Say .045

Correct Answer is: c

27. 135 Degrees F to 170 Degrees F

Correct Answer is: a

28. Using the Friction Loss Chart, 10,000 cfm in a 20 inch ϕ duct = 4,500 fpm

   Time velocity pressure is $(4,500/4,005)^2 = 1.26$ inches

   The static pressure at Point A is $6.00 - 1.26 = 4.74$ inches

   Using the Friction Loss Chart, the pressure drop is 1.2 inches per 100 feet of duct, for a total drop of 3.6 inches.

   The static pressure at Point B is $4.74 - 3.6 = 1.14$ inches

   1.14 inches of water gauge

   Correct Answer is: a

29. Install a static pressure sensor 2/3 of the way down the main duct.

   Correct Answer is: a

30. With a flexible roof diagram, loads are distributed to the resisting members based on tributary area.

   The total lateral force to the center lateral frame is 200#/ft. x 30 feet = 6,000 pounds

   The shear force in roof diagram each side of center lateral frame is 6,000/2 = 3,000 pounds

   The roof diaphragm force is $\dfrac{3,000 \; pounds}{40 \; feet}$ = 75 pounds/foot

   Correct Answer is: b

31. Column footing axial load = 100 K

   Column footing overturning moment = 50 k-ft.

   Eccentricity = M/P = 50/100 = 0.5 foot

   b/6 = 7/6 = 1.167 ft > e

   q max = P/A $(1\pm \frac{6e}{b})$ = $\dfrac{100}{7 \times 7}$ $(1\pm \frac{6 \times .5}{7})$ = 2.04 (1±.43)

   q max = 2.04 x 1.43 = 2.92 ksf
   q min = 2.04 x .57 = 1.16 ksf

   Correct Answer is: c

43.

32. Calculate moment at bottom of base plate: 10k x 34 inches = 340 k-in

Distance between centerline of anchor bolts = 12 inches

Tension in one anchor bolt = $\dfrac{340}{12 \times 2}$ = 14.2 k

Correct Answer is: b

33. Moment of inertia of shear walls is 2,190 ft.^{4}
Concrete modulus of elasticity; $Ec = 57\sqrt{f'c}$
Drift = ΔEI, Δ given at each level
Maximum inter-story drift is 0.005h

Try f'c = 5,000 psi $Ec = 57\sqrt{5,000} = 4,030$ ksi
EI = 4,030 x 2,190 x 12^4 = 183 x 10^9 k-in^2

$$\Delta_R = \frac{531 \times 10^9}{183 \times 10^9} = 2.90 \text{ inches}$$

$$\Delta_4 = \frac{403 \times 10^9}{183 \times 10^9} = 2.20 \text{ inches}$$

$$\Delta_3 = \frac{275 \times 10^9}{183 \times 10^9} = 1.50 \text{ inches}$$

$$\Delta_2 = \frac{165 \times 10^9}{183 \times 10^9} = .90 \text{ inch}$$

Check inter-story drift

4th to Roof	2.90 - 2.20	=	.70 inch
.005 x 12 x 12 = .72 inch	Okay		
3rd to 4	2.20 - 1.50	=	.70 inch
.005 x 12 x 12 = .72 inch	Okay		
2nd to 3rd	1.50 - .90	=	.60 inch
.005 x 12 x 12 = .72 inch	Okay		
1st to 2nd			.90 inch
.005 x 18 x 12 = 1.08 inches	Okay		

With 5,000 psi concrete compressive strength, the inter-story drift is less than the limit of .005 H.

With 3,000 psi and 4,000 psi concrete compressive strength, the inter-story drift exceeds the limit.

With 6,000 psi concrete compressive strength the inter-story drift is less than the limit, but is not the minimum f'c.

Correct Answer is: c

34. Calculate area of floor supported by Girder G1:

7 (8 x 3) + 8 (8 + 4) = 264 Square Feet
R = 0.08 (264 -150) = 9.12   9%

Correct Answer is: b

35. Use the provisions of the National Design Specification for Wood Construction (NDS).

Calculate the allowable bending stress:

$F'b = Fb \times C_D \times C_M \times C_T \times C_L \times C_F \times C_V$

$Fb = 1,300$ $C_D = 1.15$ $C_L = .98$ (given)

$$C_F = \frac{(12)^{1/9}}{(d)} = \frac{(12)^{1/9}}{(15.5)} = .97$$

$C_t = 1.0$

Not Applicable: C_M, C_V, Cfu, Ci, Cr, Cc, Cf

$F'b = 1,300 \times 1.15 \times .98 \times .97 = 1,421$ psi

Correct Answer is: c

36. Use the provisions of the National Design Specification for Wood Construction (NDS)

$\ell e = 1.11 \times \ell u = 1.11 \times 6 \times 12 = 79.2$

$$R_B = \sqrt{\frac{\ell e d}{b^2}} = \sqrt{\frac{79.2 \times 15.5}{5.5^2}} = 6.37 \qquad 6.4$$

Correct Answer is: a

46.

37. Calculate actual shear stress

$$Fv = \frac{3V}{2A} = \frac{3 \times 5,000}{2 \times (5.5 \times 15.5)} = 88 \text{ psi}$$

Calculate allowable shear stress

$$F'v = F_V \times C_D \times C_M \times C_t \times C_H$$

$$Fv = 140 \text{ psi} \qquad C_D = 1.15$$

$$C_t = C_H = 1.0, \quad C_M \text{ (Not Applicable)}$$
$$F'v = 140 \times 1.15 = 161 \text{ psi}$$

$$\frac{88}{161} = .55$$

Correct Answer is: c

38. Use the provisions of ACI 318, Section 7.10.5.2

Maximum Column Tie Spacing = 16db = 16 x .75 = 12 inches
or
48 d tie = 48 x .375 = 18 inches
or
least column dimension = 16 inches

12 inch controls

Correct Answer is: c

39. Support Reactions = 10k/2 = 5k

Length of member a - b = $\sqrt{4^2 + 6^2}$ = 7.21 feet

$$\frac{Force\ a - b}{7.21\ feet} = \frac{5^k}{4\ feet} \qquad \text{Force a - b} = \frac{5 \times 7.21}{4} = 9.0^k$$

or $\Sigma Fy = 0 = 5 - (Fa\text{-}b \sin 33.7)$

$$Fa - b = \frac{5}{\sin 33.7} = 9.0^k$$

Correct Answer is: d

40. Force a - b = 25^k compression

K = 1.0 ℓ = 7.21 feet   Kℓ = 7.21 feet

2-L2 x 2 x 3/8   r = .594   wt = 9.4#/f   A = 2.92in^2

$\dfrac{kl}{r} = \dfrac{7.21 \times 12}{.594}$ = 145.7   Fa = 7.01 ksi

Pall = 7.01 x 2.92 = 20.5^k N.G.

2-L2½ x 2½ x 5/16   r = .761   wt. = 10.0#/f   A = 2.93 in^2

$\dfrac{kl}{r} = \dfrac{7.21 \times 12}{.761}$ = 113.7   Fa = 11.13 ksi

Pall = 11.13 x 2.93 = 32.6^k O.K.

2-L3 x 3 x 3/16   r = .939   wt. = 7.42#/f   A = 2.18 in^2

$\dfrac{kl}{r} = \dfrac{7.21 \times 12}{.939}$ = 92.1   Fa = 13.84 ksi

Pall = 13.84 x 2.18 = 30.2^k O.K.

2-L3 x 3 x 5/16 r = .922   wt. = 12.2#/f   A = 3.55 in^2

$\dfrac{kl}{r} = \dfrac{7.21 \times 12}{.922}$ = 93.8   Fa = 13.72 ksi

Pall = 13.72 x 3.55 = 48.7^k O.K.

2-L3 x 3 x 3/16 is the lightest weight double angle which can support 25^k compression force.

Correct Answer is: c

48.

SAMPLE QUESTIONS FOR THE AFTERNOON PORTION
OF THE EXAMINATION
IN ARCHITECTURAL ENGINEERING

41. Which of the following most nearly describes how plumbing vents should be sloped?

    a. Vent lines are not required to be sloped.
    b. Plumbing vents should be sloped to drain back to the soil or waste pipe by gravity, at any amount of slope.
    c. Vent lines must be sloped toward the vent cleanout.
    d. Vent lines must be sloped a minimum of 1/4" per foot towards the soil or waste stack.

42. The term for the ratio of the Luminance leaving a surface area to the Illuminance arriving on the surface is:

    a. Exitance
    b. Transmittance Factor
    c. Luminance Factor
    d. Luminous Flux

43. Which of the following structural materials provides the greatest resistance to fire while maintaining its structural integrity?

    a. Structural steel
    b. Light wood framing
    c. Normal weight concrete
    d. Lightweight concrete

44. Considering the requirements of the UBC, BOCA, or SBC model building codes, the direction of the external design pressures on the flat roof of a building due to a horizontal wind blowing across the surface of the roof is?

    a. Upward
    b. Downward
    c. Both Upward and Downward
    d. Neither Upward nor Downward

45. In accordance with the requirements of the Americans with Disabilities Act (ADA) of 1990, which of the following toilet rooms are NOT required to meet accessibility requirements?

    a. Employee only restroom in restaurant kitchen.
    b. Toilet room with access only from a single person office.
    c. Toilet room with access only from a single exam room in doctor's office suite.
    d. Restroom on second floor of a commercial building without elevator.

46. The masonry portion of a building project consists of 22,500 brick. The crew assigned to this work is comprised of the following trades along with their base hourly cost:

    o   3 masons @ $21.00 per hour each
    o   2 apprentices @ $15.00 per hour each
    o   1 laborer @ $12.00 per hour each
    o   1 foreman @ $23.00 per hour each

    The schedule for this activity has a duration of 20 work days to complete. The labor budget associated with this activity is $15,000.00. The production rate necessary (brick per day) for this crew to achieve both the budget and the schedule is most nearly:

    a. 852
    b. 1,125
    c. 1,172
    d. 1,535

For **Question 47**, refer to the following information.

47. Wallace Jones Construction Co. was low bidder to renovate a government office complex at an Air Force Base that was being reopened. During the pre-bid site visit, each bidder was given a 20-minute tour of the facility. Because of security issues, no other investigations were allowed and no as-built drawings were available. The contract, which Jones signed, contained a standard governmental concealed conditions clause, which read in part:

Concealed Condition Clause: The contractor shall promptly, and before the conditions are disturbed, give a written notice to the contracting office of (1) Subsurface or latent physical conditions at the site that differ materially from those indicated in this contract; or (2) unknown physical conditions at the site, of an unusual nature, that differ materially from those ordinarily encountered and generally recognized as inherent in work of the character provided for in the contract.[1]

When Jones began the work, he discovered that the floor slabs were 12-inches thick instead of the 6-inches shown on the drawings which made demolition much more difficult. On the plans, the slab thicknesses were shown as approximately 6-inches with the directive that the thickness may vary and should be verified by the contractor. Jones sought additional compensation arguing that the 12-inches thick slab was a concealed condition.

The government should:

   a.   pay the contractor for the legitimate costs and time
   b.   not pay the contractor for legitimate costs and time
   c.   not pay the contractor for legitimate costs, but grant a time extension
   d.   insufficient information

48. The labor budget is $1,685.00 for a 12-inch CMU retaining wall that is 75 feet long and 6 feet high. Laborers make $8.50/hour and Masons make $17.50/hour. If your crew consists of 2 masons and 1 laborer, the maximum time (hours) allowed for completion of the retaining wall in order to meet your budget is most nearly:

   a.   32
   b.   35
   c.   39
   d.   49

For **Question 49**, refer to the following diagram.

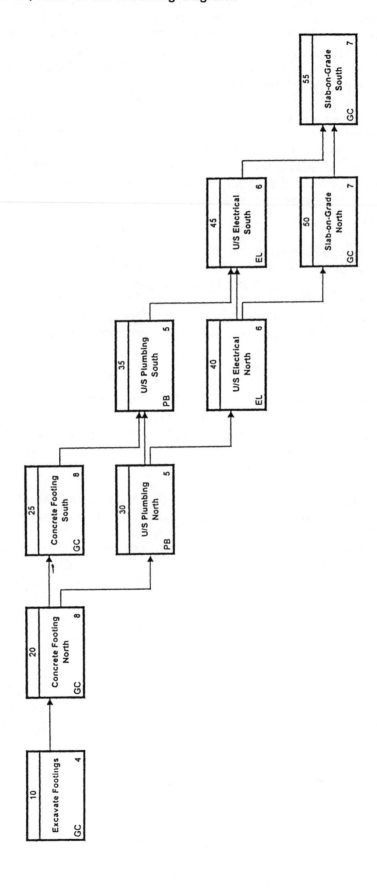

49. Using the given Precedence Method CPM logic diagram for the construction of a slab on grade. The Activity Duration is shown on the lower right hand corner of the rectangle. The activity number is listed at the top of the rectangle. The activity description is listed in the middle. The responsibility code is listed in the lower left hand corner. For example, Activity #25 Concrete Footing for the South Half of the Building will take 8 days and will be done by the General Contractor (GC).

For the given Precedence Method CPM logic diagram, the total duration (days) necessary to construct the slab on grade is:

a. 35
b. 36
c. 38
d. 40

For **Question 50**, refer to the following Payment Application.

Building Renovation Project

PAYMENT APPLICATION NUMBER: 2

DESCRIPTION OF WORK	SCHEDULED VALUE	WORK COMPLETED			TOTAL COMPLETED AND STORED TO DATE	%	BALANCE TO FINISH	RETAINAGE
		PREVIOUS APPLICATIONS	THIS APPLICATION					
			WORK IN PLACE	STORED MATERIAL				
GENERAL REQUIREMENTS	70,000.00	40,000.00	10,000.00	0.00	50,000.00	71%	20,000.00	5,000.00
DEMOLITION/SITEWORK	12,000.00	12,000.00	0.00	0.00	12,000.00	100%	0.00	1,200.00
CONCRETE	10,000.00	10,000.00	0.00	0.00	10,000.00	100%	0.00	1,000.00
MASONRY	10,000.00	7,500.00	2,500.00	0.00	10,000.00	100%	0.00	1,000.00
ROUGH & FINISH CARPENTRY	20,000.00	0.00	1,500.00	3,000.00	4,500.00	23%	15,500.00	450.00
CAULKING & SEALANTS	4,500.00	3,500.00	1,000.00	0.00	4,500.00	100%	0.00	450.00
DOOR, FRAMES & HARDWARE	25,000.00	0.00	2,500.00	7,500.00				
WINDOWS, GLASS & GLAZING	9,000.00	4,500.00	4,500.00	0.00	9,000.00	100%	0.00	900.00
DRYWALL & CEILINGS	35,000.00	20,000.00	15,000.00	0.00				
FLOORING, VCT & CARPET	10,500.00	0.00	0.00	0.00	0.00	0%	10,500.00	0.00
SPECIAL COATINGS	4,500.00	0.00	0.00	0.00	0.00	0%	4,500.00	0.00
PAINTING	12,000.00	4,000.00	4,000.00	0.00	8,000.00	67%	4,000.00	800.00
SPECIALTIES	9,000.00	0.00	0.00	0.00	0.00	0%	9,000.00	0.00
CASEWORK	12,000.00	0.00	0.00	0.00	0.00	0%	12,000.00	0.00
PLUMBING/HVAC	110,000.00	90,000.00	10,000.00	0.00	100,000.00	91%	10,000.00	10,000.00
FIRE PROTECTION PIPING	14,500.00	14,500.00	0.00	0.00	14,500.00	100%	0.00	1,450.00
ELECTRICAL	90,000.00	70,000.00	10,000.00	0.00	80,000.00	89%	10,000.00	8,000.00
TOTAL	458,000.00	276,000.00	61,000.00	10,500.00	347,500.00	76%	110,500.00	34,750.00

50. The contractor has submitted the given Payment Application No. 2 for the "Building Renovation Project". The amount of retention withheld from progress payments is 10%. The amount that should be paid this month for Drywall & Ceilings is most nearly: (Assume all previous payments have been paid)

    a. $11,500
    b. $13,500
    c. $15,000
    d. $31,500

51. Construction Management Project Delivery Methods include Construction Manager - Advisor Contracts and Construction Manager - At Risk Contracts. The Construction Manager - At Risk is liable to the Owner for:

    a. Design Errors & Omissions
    b. Differing Site Conditions
    c. Force Majeure Weather Delays
    d. Contractor Defaults

52. A 208-volt, 3-phase, 4-wire, 600-ampere continuous linear load is to be fed through a transformer from an existing 480-volt, 3-phase, 4-wire switchboard.

    Using the given information about this installation, the minimum standard size transformer (kVA) that can be used to serve this load is most nearly: (Assume the transformer is 100% rated.)

    a. 150
    b. 225
    c. 300
    d. 500

For **Question 53**, refer to the following diagram.

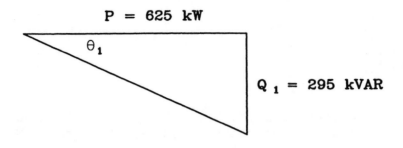

P = 625 kW

θ_1

Q_1 = 295 kVAR

53. For the power triangle shown in the given diagram, the amount of capacitance in kVAR that must be added to bring the power factor to .95 is most nearly:

a. 90
b. 205
c. 295
d. 691

54. A factory has the following loads:

1. 60 kilowatts of resistance heating
2. 100 kilowatts at a Lagging Power Factor = 0.707
3. 40 kilowatts of Leading Power Factor = 0.80

The total kVA and the overall power factor for these loads is most nearly:

a. 203.7 kVA @ .982 Lagging
b. 203.7 kVA @ .982 Leading
c. 211.9 kVA @ .944 Lagging
d. 211.9 kVA @ .944 Leading

For **Question 55**, refer to the following diagrams.

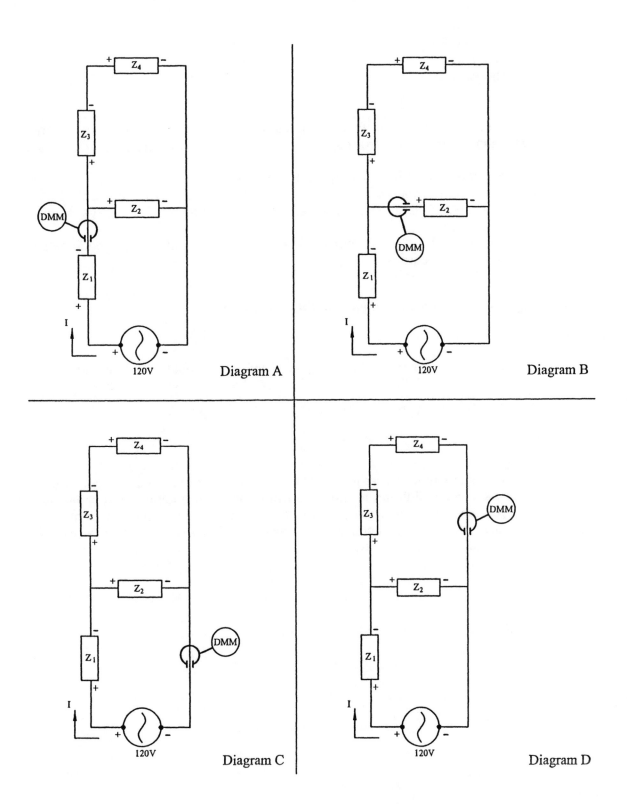

Diagram A

Diagram B

Diagram C

Diagram D

55. Using the given diagrams, the correct location of a clamp-on digital multimeter (DMM) to measure the current flowing in load Z_2 is shown in?

   a. Diagram A
   b. Diagram B
   c. Diagram C
   d. Diagram D

56. In constructing a Single Phase, Series Vector Diagram, which one of the following statements is true?

   a. The reference axis is always drawn vertical.
   b. The resistive component and reference axis are always "out of phase."
   c. The phase relationship between vector components is always counter clockwise.
   d. The phase relationship between the impedance and the reference axis is always 90 degrees.

For **Question 57**, refer to the following information.

Lamp	Input Volts	Ballast Type	Input Watts	Input Current	Open Ckt. Voltage
H37LC-250/C	120	HX-NPF	285	4.80	240

57. A convenience store owner in a rural area is purchasing mercury vapor "yard" lights as security lighting around the property. The equipment supplier has provided the given operating data for the ballast. The MAXIMUM number of luminaires (yard lights), according to the NEC, that can be put on a 20A1P circuit breaker is most nearly:

   a. 8
   b. 6
   c. 3
   d. 2

60.

For **Question 58**, refer to the following Room Geometry and Photometric Data.

ROOM B—GEOMETRY:

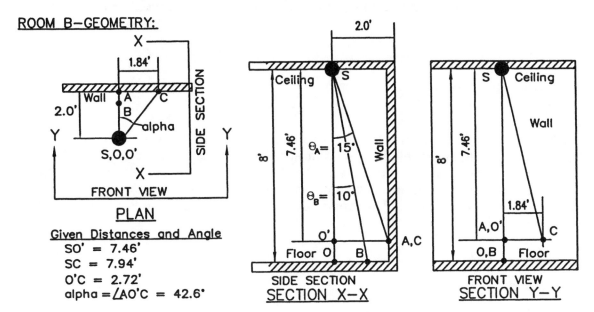

PLAN

Given Distances and Angle
SO' = 7.46'
SC = 7.94'
O'C = 2.72'
alpha = ∠AO'C = 42.6°

SIDE SECTION
SECTION X—X

FRONT VIEW
SECTION Y—Y

PHOTOMETRIC DATA

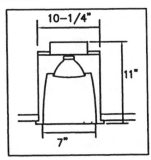

Luminaire —"BAE150B"

DESCRIPTION:
7" Recessed Incandescent
Downlight with Clear
Low Brightness Cone

LAMP: 150PAR38/FL VOLTS 120
LUMEN/LAMP 1740 DATE 1993
SHIELDING ANGLE: N 40 L 40
BARE 38000 FILE 100—R3
TEST NO. 72703 BY: RNH

CATEGORY IV

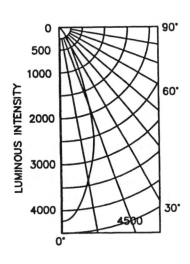

Luminaire Spacing Criterion = .60

LUMINOUS INTENSITY

θ	I
5	3987
10	3387
15	2786
20	1881
25	974
30	524
35	74
40	65
45	55
50	0
55	0

61.

58. Using the given Room Geometry and Photometric Data, the illuminance level, (fc), E_A, at point "A" on the vertical surface from the source S, is most nearly:

a. 12.1
b. 13.0
c. 45.2
d. 48.4

59. The utility serving the building you are designing the power distribution tells you that the available fault current at the service entrance is 22 MVA at 480 volts. The minimum allowable interrupting rating (A) of the main circuit breaker at your service entrance that meets the requirements of the National Electrical Code is most nearly: The ratings (A) listed below are "standard" interrupting ratings.

a. 22,000
b. 42,000
c. 65,000
d. 100,000

60. Using the Energy Code, ASHRAE/IES90.1-1989, determine which of the following statements apply to the given room. The room is a "Library" room that has the dimensions of 30 feet by 45 feet with an Area Factor, AF = 1.15. Two visual tasks occur in this room which are (1) card filing that is 20% of the area and has a UPD = 1.6 and (2) a reading area that is 80% of the area and has a UPD = 1.9.

I.   The lighting power allowance should be calculated using a UPD = 1.84
II. The lighting power allowance should be calculated using a UPD = 1.75.
III. The LPB for this room is most nearly 2,857 watts.
IV. The LPB for this room is most nearly 2,717 watts.

a. I and III only
b. II and III only
c. I and IV only
d. II and IV only

61. A 6-inch pump operating at 1,800 rpm discharges 1,800 gpm of cold water (S.G. = 1) against an 80 foot head at 60% efficiency. A dynamically similar 8-inch pump operating at 1,170 rpm is considered as a replacement. The total head (ft.) that can be expected from the new pump is most nearly:

a. 52.00
b. 60.09
c. 106.15
d. 106.66

62. According to the requirements of NFPA 13-1999, the maximum floor area (ft^2) on any one floor to be protected by sprinklers supplied by any one sprinkler system riser or combined system riser for ordinary hazard occupancy should be most nearly:

    a.   25,000
    b.   52,000
    c.   60,000
    d.   Unlimited floor area

63. With a chilled water and hot water primary system, which of the following air distribution systems delivers supply air from the air handling unit at a constant temperature during summertime operation:

    a.   A system consisting of fan-coil units.
    b.   A system consisting of single zone air handling units.
    c.   A system consisting of variable air volume (VAV) air handling units.
    d.   A system consisting of constant volume, double duct air handling units.

For **Question 64**, use the psychometric chart, below.

Copyright 1992 by the American Society of Heating, Refrigerating, and Air-Conditioning Engineers, Inc.
Used by permission.

ASHRAE PSYCHROMETRIC CHART NO. 1

NORMAL TEMPERATURE

BAROMETRIC PRESSURE: 29.921 INCHES OF MERCURY

COPYRIGHT 1992

AMERICAN SOCIETY OF HEATING, REFRIGERATING AND AIR-CONDITIONING ENGINEERS, INC.

SEA LEVEL

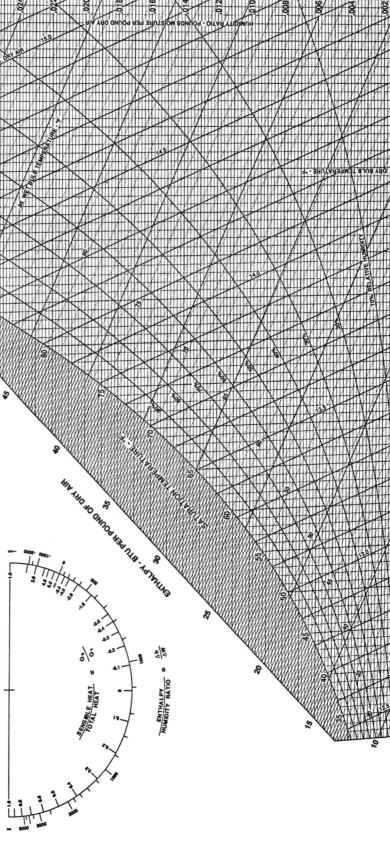

Prepared by: CENTER FOR APPLIED THERMODYNAMIC STUDIES, University of Idaho

64. An air handling unit circulates a total of 20,000 CFM, of which 2,000 CFM is outside air. Given a return air condition of 80 degrees F, 60% RH and an outside air condition of 95 degrees F, 50% RH, the mixed air condition (degrees F, % RH) is most nearly:

    a. 93.6 degrees F and 51% RH
    b. 87.5 degrees F and 56% RH
    c. 85.0 degrees F and 56% RH
    d. 81.4 degrees F and 59% RH

65. For which of the flowing systems must the elevation head be taken into consideration when selecting a pump?

    a. A condenser loop for an open cooling tower.
    b. A chilled water pump for an HVAC system.
    c. A hot water pump for a hydronic heating system.
    d. A domestic hot water re-circulation pump.

66. A condensing unit contains:

    a. Condenser fan, condenser coil and compressor
    b. Condenser fan and condenser coil
    c. Evaporative condenser fan and condenser coil
    d. Condenser fan, condenser coil, compressor, evaporator fan and evaporator coil

67. A fan motor consumes 40 kilowatts of power, operates 2,080 hours per year and delivers 35,000 CFM (constant volume). The air distribution system is converted to variable air volume and the same fan supplies an annual average of 20,000 CFM at the same static pressure. The annual amount of energy saved, assuming $0.08 kW-h, is most nearly:

    a. $6,650
    b. $5,420
    c. $4,480
    d. $2,850

For **Question 68**, use the psychometric chart, below.

Copyright 1992 by the American Society of Heating, Refrigerating, and Air-Conditioning Engineers, Inc. Used by permission.

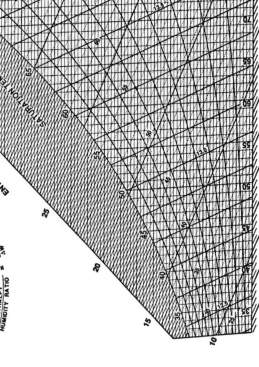

ENTHALPY - BTU PER POUND OF DRY AIR

68. The entering and leaving air conditions for a 5,000 cfm cooling coil are: 80 degrees F, 60%RH, and 55 degrees F, 95%RH. Assuming a 10 degrees F temperature difference across the coil, the volume of chilled water (gpm) that must be supplied to the coil to satisfy these conditions is most nearly:

    a. 12
    b. 50
    c. 105
    d. 250

For **Question 69**, refer to the following performance curve.

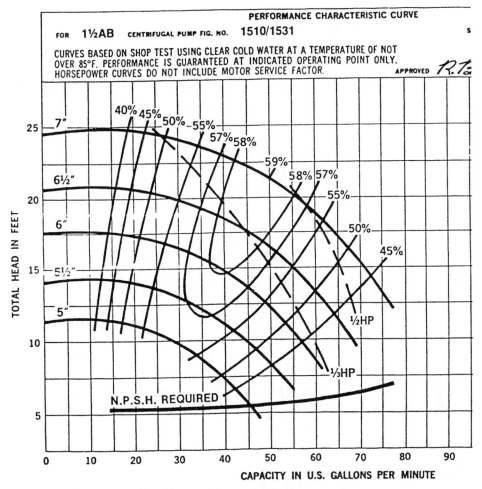

PERFORMANCE CHARACTERISTIC CURVE

FOR **1½AB** CENTRIFUGAL PUMP FIG. NO. **1510/1531**

CURVES BASED ON SHOP TEST USING CLEAR COLD WATER AT A TEMPERATURE OF NOT OVER 85°F. PERFORMANCE IS GUARANTEED AT INDICATED OPERATING POINT ONLY. HORSEPOWER CURVES DO NOT INCLUDE MOTOR SERVICE FACTOR.   APPROVED

Used by permission of Bell and Gossett.

69. Using the given performance curve, 17.5 feet of head could be achieved with a flow rate of 45 GPM and a motor of 1/3 HP.   If the 17.5 feet of head is constant, the flow rate (gpm) that could be achieved if the motor horsepower was increased to 1/2 HP is most nearly: (Assume no change in impeller size or RPM.)

    a. 45
    b. 55
    c. 65
    d. 70

For **Question 70**, refer to the following sketch.

Footing Concrete Compressive Strength = 3000 psi

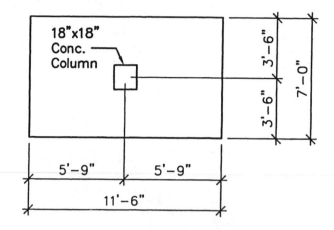

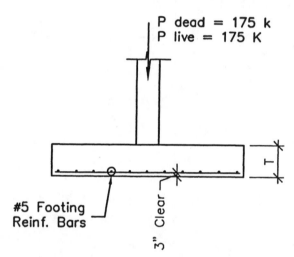

70. Using the given column footing and column loading, the minimum thickness (inches) of the column footing to resist punching shear is most nearly:

    a. 20
    b. 24
    c. 32
    d. 35

For **Question 71**, refer to the following figures.

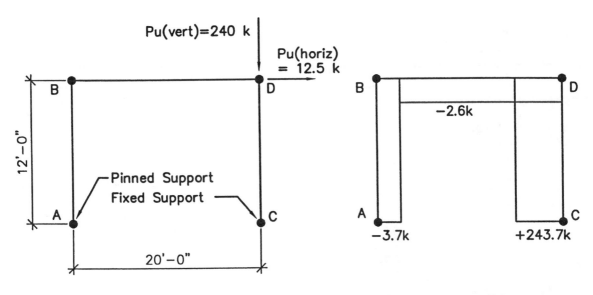

Pu(vert)=240 k

Pu(horiz) = 12.5 k

B

D

12'-0"

Pinned Support
Fixed Support

A

C

20'-0"

Figure 1

B   D

−2.6k

A   C

−3.7k   +243.7k

Axial Load Diagram

Figure 2

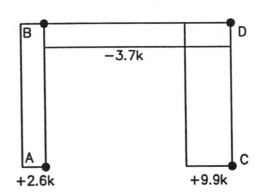

B   D

−3.7k

A   C

+2.6k   +9.9k

Shear Diagram

Figure 3

71. Given a rigid frame and factored loading as shown on Figure 1, the axial load diagram for the rigid frame as shown on Figure 2, and the shear diagram for the rigid frame as shown on Figure 3, the most probable moment diagram of the following moment diagrams is most nearly:

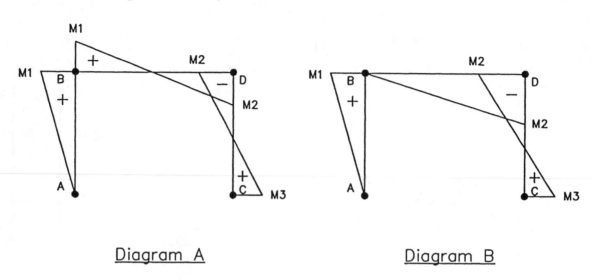

Diagram A   Diagram B

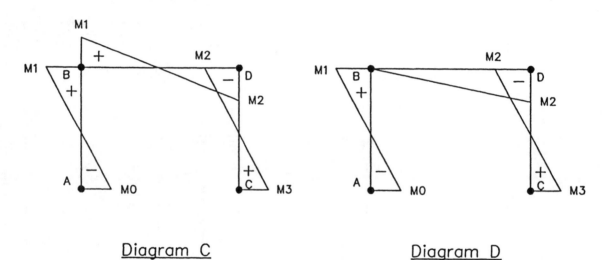

Diagram C   Diagram D

a. Diagram A
b. Diagram B
c. Diagram C
d. Diagram D

72. The minimum length (inches) of a 1/4 inch E70xx electrode fillet weld required to support a dead plus live load of 20 kips is most nearly:

a. 1.1
b. 1.6
c. 3.8
d. 5.4

For **Question 73**, refer to the following section and information.

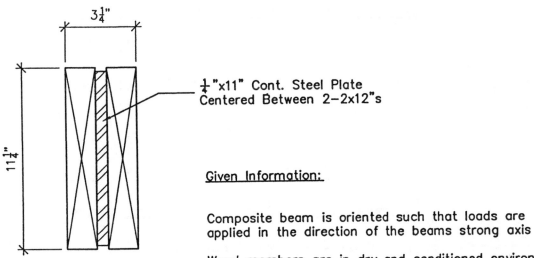

¼"x11" Cont. Steel Plate
Centered Between 2-2x12"s

Composite Beam Section

Given Information:

Composite beam is oriented such that loads are applied in the direction of the beams strong axis

Wood members are in dry and conditioned environment

Wood: E = 1,350,000 psi

Steel: Fy = 36 ksi

Steel: E = 29,000,000 psi

73. Using the given cross-section and information, a steel plate is sandwiched between two 2x12's and the members are bolted together so the steel plate and wood member act as one unit. The transformed wood moment of inertia (in⁴) for the given section is most nearly?

    a. 356
    b. 384
    c. 952
    d. 7,680

For **Question 74**, refer to the following roof framing plan.

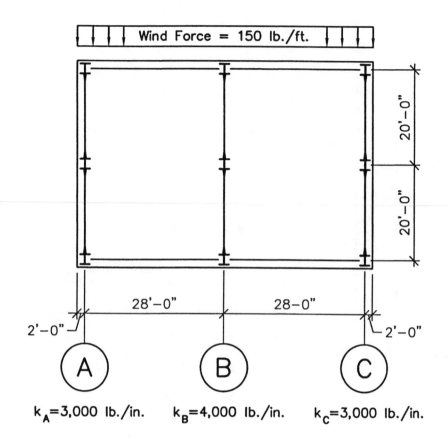

74. Using the given roof framing plan, wind force loading at the roof level, and stiffness of each frame, the maximum required capacity of the rigid diaphragm (lb./ft.) is most nearly:

    a.   60
    b.   63
    c.   68
    d.   90

For **Questions 75-76**, refer to the following subsurface soil condition and geotechnical properties.

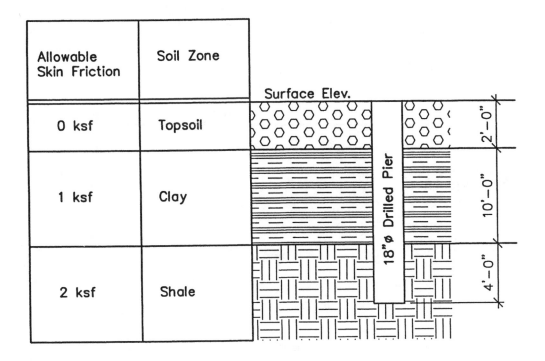

Allowable Skin Friction	Soil Zone
0 ksf	Topsoil
1 ksf	Clay
2 ksf	Shale

End Bearing Capacity = 4.0 ksf

Figure 1

75. Using the given subsurface soil conditions and geotechnical properties shown in Figure 1, the maximum allowable capacity (kips) of the 18" diameter straight shaft drilled pier is most nearly:

    a. 73
    b. 92
    c. 103
    d. 120

76. Using the given subsurface soil conditions and geotechnical properties shown in Figure 1, and the shaft length within the clay layer of 10 feet, the required penetration (feet) into the shale to develop the required pier capacity of 120 kips is most nearly:

    a. 4.0
    b. 7.0
    c. 11.5
    d. 14.0

73.

For **Question 77**, refer to the following Beam Loading Diagram and information.

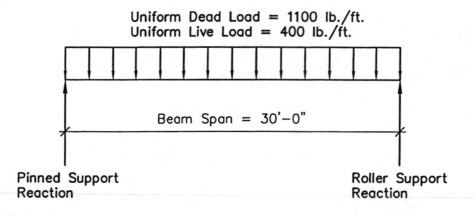

Uniform Dead Load = 1100 lb./ft.
Uniform Live Load = 400 lb./ft.

Beam Span = 30'-0"

Pinned Support
Reaction

Roller Support
Reaction

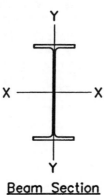

Beam Section

E = 29,000 ksi
Fy = 36 ksi

77. Using the given beam span, loading, and properties, the minimum moment of inertia (in⁴) about the x-axis of the given beam to limit the mid-span deflection to 1/2" when using ONLY 75% of dead load is most nearly:

a. 100
b. 375
c. 1,040
d. 1,400

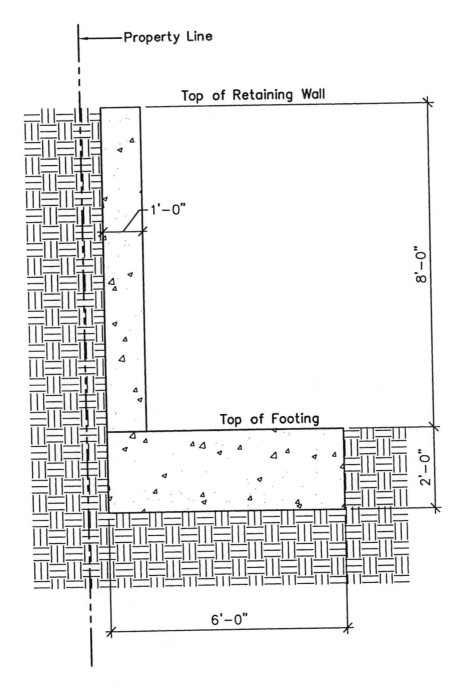

For **Questions 78**, refer to the following details.

Property Line

Top of Retaining Wall

1'-0"

8'-0"

Top of Footing

2'-0"

6'-0"

Retaining Wall Section — Figure 1

78. Using the given retaining wall geometry and details in Figure 1, and a 1 foot unit width of the retaining wall, the unfactored moment (lb.-ft.) at the top of the footing due to the lateral earth pressure load of 40 pcf equivalent fluid pressure is most nearly:

a. 3,400
b. 6,600
c. 6,800
d. 10,000

75.

For **Question 79**, use the following diagram and information:

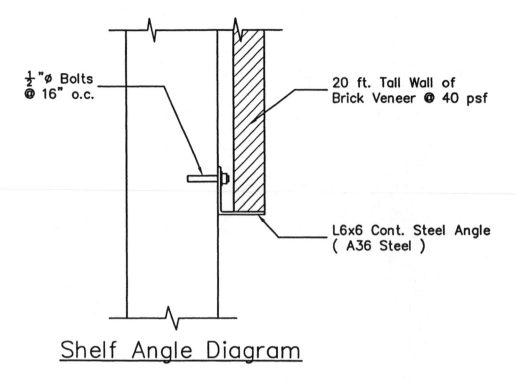

$\frac{1}{2}$ "∅ Bolts
@ 16" o.c.

20 ft. Tall Wall of
Brick Veneer @ 40 psf

L6x6 Cont. Steel Angle
(A36 Steel)

Shelf Angle Diagram

79. Using the given shelf angle diagram and the given information, the shear force on the bolt (pounds) from the weight of the brick is most nearly:

   a. 600
   b. 800
   c. 1,100
   d. 1,200

For **Question 80**, refer to the following information.

A mezzanine floor in a warehouse spans 20 feet as a single simple span. The floor is a reinforced concrete one-way slab, 12 inches thick. There are no other dead loads, but non-structural elements that are likely to be damaged by large deflections will be attached to the slab 28 days after the slab concrete is placed.

GIVEN INFORMATION:

Concrete : $f'c$ = 4,000 psi (normal weight) weight = 150 lbs./cu. ft.
Reinforcing Steel : fy = 60,000 psi
Service Live Load : 250 psf

Code: 1995 or 1999 Edition of ACI 318, Building Code Requirements for Structural Concrete

80. Assuming the full service live load is sustained, I_e = 670 in.^{4}/ft., λ = 1.4 and the non-structural elements are attached after the shores are removed, the maximum computed long-term deflection of the slab (inches) that will occur under full service load after the non-structural elements are attached is most nearly:

   a. 0.17
   b. 0.60
   c. 0.85
   d. 1.33

SOLUTIONS FOR SAMPLE QUESTIONS
FOR THE AFTERNOON PORTION
OF THE EXAMINATION
IN ARCHITECTURAL ENGINEERING

Solutions for Sample Questions
for the Afternoon Portion
of the Examination
in Architectural Engineering

41. Plumbing vents should be sloped to drain back to the soil or waste pipe by gravity, at any amount of slope.

    Correct Answer is: b

42. Luminance Factor

    Correct Answer is: c

43. Lightweight concrete

    Correct Answer is: d

44. Upward

    Correct Answer is: a

45. In accordance with ADAAG 4.1.3 (11), a toilet room with access only from a single person's office does not need to meet accessibility requirements.

    Correct Answer is: b

46. Based upon allowed duration: 22,500/20 = 1125 bricks/day

    Based upon allowed costs:

    Crew cost per hour = (3x21) + (2x15) + (1x12) + (1x23) = $128.00

    15,000/128x8 = 14.65 days

    22,500/14.65 = 1,535 brick/day

    Budget controls, therefore the necessary rate is 1,535 brick/day

    Correct Answer is: d

47. While the plans did not speak with certainty as to the slab thickness and the contractor was warned that the slab thickness may vary, these disclaimers cannot overcome the fact that a 20 minute site tour is insufficient time for a contractor to verify all the aspects needed to formulate an accurate bid. The contractor was justified in relying on the information provided by the government. The government should pay the contractor for legitimate costs and time.

Correct Answer is: a

48. 2 masons @ $17.50/hr   = $35.00
     1 laborer @ $8.50/hr   = $8.50
     Total Labor Cost   = $43.50 /hr.
     $1,685.00/43.50   =   38.8 hours

     39 Hours

     Correct Answer is: c

49. The total duration necessary to construct the slab on grade is calculated as the longest path through the network.

Activities	:	10	-	Excavate Footings	:	4 days
		20	-	Concrete Footings North	:	8 days
		25	-	Concrete Footings South	:	8 days
		35	-	Underslab Plumbing South	:	5 days
		45	-	Underslab Electrical South	:	6 days
		55	-	Slab on Grade South	:	7 days
						38 days

     Correct Answer is: c

50.

Total Completed and Stored to Date Through Payment Application No. 2	$35,000.00
Total paid on previous Application	$20,000.00
Amount Due before Retention	$15,000.00

Total Retention Withheld through Payment Application No. 2	$ 3,500.00
Amount of Previous Retention from Payment Application No. 1 10% x 20,000.00 =	$ 2,000.00

Amount of Retention withheld from Payment Application, No. 2	$ 1,500.00
Amount Paid this Month for Drywall and Ceilings (15,000 - 1,500)	$13,500.00

Correct Answer is: b

51.   Contractor Defaults

If a contractor defaulted that was under contract to a CM at Risk, the CM would become responsible to complete the work. If the contractor was bonded, the CM may take action against the bond to try and force the bonding company to step in and complete the work of the defaulted contractor for the CM. A Construction Manager - at Risk holds contracts. A Construction Manager - Advisor does not hold any contracts. All contracts are prime with the Owner.

Correct Answer is: d

52. $\sqrt{3 \times} \dfrac{208\,Volts \times 600\,amps}{1,000} = 216.2 \text{ kVA}$

The minimum standard size transformer is 225 kVA

Correct Answer is: b

53. Current Power Factor, $\cos\theta = \dfrac{P}{\sqrt{P^2 + O^2}}$

   $\cos\theta = \dfrac{625}{\sqrt{625^2 + 295^2}} = .90$

   PF $= .90$
   PF $= .95$ $.95$ theta $= \cos - 1\,(95) = 18.19$

   $Q_2 = 625 \times \tan 18.19 = 205.37$, say 205
   $Q_1 - Q_2 = 295 - 205 = 90$ kVAR

   Correct Answer is: a

---

54. 60 kW + 100 kW + 40 kW = 200 kW
   theta $= \cos - 1\,(.707) = 45°$
   VARS $= 100 \times \tan 45 = 100$ Lagging
   theta $= \cos - 1\,(.80) = 36.87°$
   VARS $= 40 \times \tan 36.87 = 30$ Leading
   $\sum$ VARS $= 100 - 30 = 70$ Lagging

   kVA $= \sqrt{200^2 + 70^2} = 211.9$
   theata $= \tan - 1\,(70/200) = 19.29$
   PF $= \cos 19.29 = .944$
   211.9 kVA @ .944 Lagging

   Correct Answer is: c

55. Diagram B

   Correct Answer is: b

56. The rotation of vectors is always assumed counter clockwise.

   Correct Answer is: c

57. 20A1P = 120 x 20 x 80% = 1,920 VA, (maximum by code)
   1,920/120x4.80 = 3.3   3 luminaries

   Correct Answer is: c

58. $\theta_A = 15°$

$D = 7.46/\cos 15° = 7.72'$

$E_A = \dfrac{I_A}{D^2} \times \sin \theta_A = \dfrac{27.86}{(7.72)^2} \times \sin 15° = 12.09$

$E_A = 12.1 \ fc$

Correct Answer is: a

59. $I_{fault} = \dfrac{22 \ MVA}{\sqrt{3} \times 480 \ volts} \times (1 \times 10^6) = 26{,}462 \ Amps$

Next largest is 42,000 Amps

Correct Answer is: b

60. UPD avg. = $[(1.6 \times 20) + (1.9 \times 80)]/100 = 1.84$

LPB = $1.84 \times 1.15 \times (30 \times 45) = 2{,}587$ watts

I & III only

Correct Answer is: a

61. H = $(8 \times 1170/6 \times 1800)^2 \times 80 = 60.09$ feet

Correct Answer is: b

62. According to NFPA 13 (1999) 4.2.2.1, the maximum floor area for an ordinary hazard is 52,000 Square Feet.

Correct Answer is: b

63. A system consisting of variable air volume (VAV) air handling units.

Correct Answer is: c

64.   The specific volume of the return air is approximately 13.88 cubic feet per pound of dry air. The mass flow rate is 18,000/13.88 = 1,297 lbm per minute. The specific volume of the outside air is approximately 14.36 cubic feet per pound. The mass flow rate is 2,000/14.36 = 139 lbm per minute.

The total mass flow rate is 1,297 + 139 = 1,436 lbm per minute.

Calculating the mixed air temperature:

$$
\begin{array}{rl}
1{,}297 \times 80/1436 = & 72.2 \text{ degrees} \\
139 \times 95/1436 = & \underline{9.2 \text{ degrees}} \\
& 81.4 \text{ degrees}
\end{array}
$$

At 81.4 degrees on the mixing line on the psychrometric chart, the humidity is 59% RH

81.4° degrees and 59% RH

Correct Answer is: d

65.   A condenser loop for an open cooling tower.

Correct Answer is: a

66.   A condensing unit contains a condenser fan, condenser coil, and compressor.

Correct Answer is: a

67.   In the original operating mode, the fan consumes:

40 kW x 2,080 hours x 0.08 = $6,656.00

Using the fan laws:

$$ W_1 = W_2 \times (Q_1/Q_2)^3 = 40 \times \left(\frac{20{,}000}{35{,}000}\right)^3 = 7.46 \text{ kW} $$

The new annual operating cost is: 7.46 x 2,080 x 0.08 = $1,241.00

The annual savings = $6,656.00 - $1,241.00 = $5,417.00

Say $5,420.00

Correct Answer is: b

86.

68. The enthalpy of the entering air = 33.6 BTU/lb;
    The enthalpy of the leaving air = 22.6 BTU/lb.;
    33.6 - 22.6 = 11.0 BTU/lb.

    The heat absorbed by the coils = 4.5 x 5,000 x 11.0 = 247,500 BTU/hr.

    The heat absorbed by the water = 247,500 = Gallons/hr x 8.34 lbs./gal. x 10°F
    Gallons per hour = 2,967
    Gallons per minute = 49.5; say 50 GPM

    Correct Answer is: b

69. 17.5 foot of head would be the maximum at the conditions indicated. With a 1/2
    hp motor, the flow rate would not change. 45 GPM

    Correct Answer is: a

70. $\phi Vc \geq Pu$ $Vc = Pu/\phi$ $\phi = .85$
    $Pu = 1.4 \times 175 + 1.7 \times 175 = 542.5 \, k$
    $Pu/\phi = 542.5/.85 = 638.2^k$

    $Vc = 4\sqrt{f'c} \times d \times b_o = .219 \times d \times b_o$

    $d = t-4"$ $b_o = (18+d) \, 4$
    $638.2 = .219 \times d \times [(18+d) \times 4]$
    $638.2 = 15.77 \, d + .876 \, d^2$ $d = 19.5$ inches
    $t = 19.5 + 4 = 23.5$ inches, say 24 in.

    Correct Answer is: b

71. $M_1 = +2.6^k \, (12 \, ft.) = +31.2 \, k\text{-}ft.$
    $M_2 = M_1 - (3.7^k \times 20 \, ft) = 31.2 - 74 = -42.8 \, k\text{-}ft.$
    $M_3 = M_2 + (9.9^k \times 12 \, ft.) = -42.8 + 118.8 = +76.0 \, k\text{-}ft.$

    Moment Diagram A

    Correct Answer is: a

72. Allowable weld stress = 0.3 x 70 = 21 ksi

Effective weld thickness = .707 x .25 = .177 inch

Weld strength = .177 x 21 = 3.71 k/in.

Weld length required = $\dfrac{20}{3.71}$ = 5.4 inch

Correct Answer is: d

73.   n = $\dfrac{E\ steel}{E\ wood}$ = $\dfrac{29,000,000}{1,350,000}$ = 21.5

I = $\dfrac{bh^3}{12}$ (wood) + $\dfrac{bh^3}{12}$ (steel) x n

$\dfrac{3(11.25)^3}{12}$ + $\dfrac{.25(11)^3}{12}$ (21.5)

356 + 596 = 952 in.^{4}

Correct Answer is: c

74. A rigid diaphragm means that the load distribution is based on relative stiffness.

$\sum F$ = 150 lb/f x 60 ft. = 9,000 pounds

$\sum$ Stiffness = 3,000 + 4,000 + 3,000 = 10,000 lb./in.

F_A = 3/10 x 9,000 = 2,700 pounds

F_B = 4/10 x 9,000 = 3,600 pounds

F_C = 3/10 x 9,000 = 2700 pounds

Maximum Shear force in diaphragm = 2,700 - 2 (150) = 2,400 lbs.

Maximum required diaphragm capacity = $\dfrac{2,400}{40}$ = 60 lb./ft.

Correct Answer is: a

88.

75. End bearing area of 18 inch ϕ pier = 1.77 ft.^{2}
Circumference of 18 inch ϕ pier = 4.71 ft/lin. ft.

Calculate end bearing capacity = 1.77 x 4.0 = 7.1k
Calculate skin friction capacity in clay = 4.71 x 1 x 10 = 47.1k
Calculate skin friction capacity in shale = 4.71 x 2 x 4 = 37.7k
     Total pier capacity   = 91.9k

Correct Answer is: b

76. End bearing area of 18 inch ϕ pier = 1.77 ft.^{2}
Circumference of 18 inch ϕ pier = 4.71 ft.^{2}/lin. ft.

Calculate end bearing capacity = 1.77 x 4.0 = 7.1k
Calculate skin friction capacity in clay = 4.71 x 1 x 10 = 47.1 k
                                  Σ = 54.2 k

Required friction capacity in shale = 120.0 - 54.2 = 65.8k

Calc. required penetration into shale = $\dfrac{65.8}{4.71 \times 2}$ = 6.98 feet

Correct Answer is: b

77. $\Delta_{DL} = \dfrac{5 \times W_{DL} \times \ell^4}{384 \times E \times I} = 0.5"$

Calc. I required for ½″ deflection when using only 75% of dead load.

$$I = \dfrac{5 \times (1.1 \times .75) \times (30)^4 \times (1728)}{384 \times 29,000 \times .5} = 1,037 \text{ in.}^4$$

Correct Answer is: c

78. Equivalent fluid pressure of 40 pcf has a triangular distribution

Moment at top of footing $= 40 \times 8 \times \dfrac{8}{2} \times \dfrac{8}{3} = 3,413$ # - ft./ft.

Correct Answer is: a

89.

79. Weight of brick = 20' x 40 psf = 800 lbs./ft.

Anchors at 16" on center.

Shear Force per bolt = 800 x 1.33 = 1,064 pounds, say 1,100 pounds.

Correct Answer is: c

80. $E_c = \dfrac{57,000\sqrt{f'c}}{1,000} = \dfrac{57,000\sqrt{4,000}}{1,000} = 3,605$ ksi

$\Delta = \dfrac{5w\ell^4}{384\ E_c I_e} = \dfrac{5 \times .400 \times (20)^4 \times (1,728)}{384 \times 3,605 \times 670} = 0.596$ inch

Long term Δ = 1.4 x 0.596 = 0.834 inch, say 0.85"

Correct Answer is: c

REFERENCES

91.

REFERENCE MATERIAL

Occupational Safety and Health Standards for Construction Industry, 29 CFR Part 1926, US Department of Labor, latest edition

American Concrete Institute SP-4, Formwork for Concrete, 6th Edition.

Engineering Economy text book containing equations and tables related to time value of money.

Uniform Building Code, 1997 Edition, International Conference of Building Officials.

The BOCA National Building Code, 1999 Edition, Building Officials and Code Administrators International Inc.

Standard Building Code, 1999 Edition, Southern Building Code Congress International, Inc.

International Building Code, 2000 Edition, International Code Council, Inc.

ASCE 7, Minimum Design Loads for Buildings and Other Structures, 1998 Edition, American Society of Civil Engineers.

ASHRAE Fundamentals Handbook 2001, American Society of Heating, Refrigeration, & Air-Conditioning Engineers

ASHRAE HVAC Applications Handbook 1999, American Society of Heating, Refrigeration, & Air-Conditioning Engineers

ASHRAE HVAC Systems & Equipment Handbook 2000, American Society of Heating, Refrigeration, & Air-Conditioning Engineers

ASHRAE Refrigeration Handbook 2002, American Society of Heating, Refrigeration, & Air-Conditioning Engineers

ASHRAE Principles of Heating, Ventilating and Air-Conditioning, American Society of Heating, Refrigeration, & Air-Conditioning Engineers

ASHRAE Standard 90.1-2001 - Energy Standard for Building Except Low-Rise Residential Buildings, American Society of Heating, Refrigeration, & Air-Conditioning Engineers

IEEE Color Books Set 2001, Institute of Electrical and Electronic Engineers

IESNA Lighting Handbook 9th Edition, Illuminating Engineering Society of North America

SFPE Handbook of Fire Protection Engineering, 3rd Edition, Society of Fire Protection Engineers

ASPE Data Book Volume 1: Fundamentals of Plumbing Engineering, American Society of Plumbing Engineers

ASPE Data Book Volume 2: Plumbing Systems, American Society of Plumbing Engineers

ASPE Data Book Volume 3: Special Plumbing Systems, American Society of Plumbing Engineers

NFPA 70 National Electric Code 2002 Edition, National Fire Protection Association

NFPA 13 Standard for the Installation of Sprinkler Systems 1999 Edition, National Fire Protection Association

NFPA 14 Standard for the Installation of Standpipe, Private Hydrant, and Hose Systems 2000 Edition, National Fire Protection Association

NFPA 72 National Fire Alarm Code 1999 Edition, National Fire Protection Association

NFPA 99 Standard for Health Care Facilities 2002 Edition, National Fire Protection Association

NFPA 101 Life Safety Code 2000 Edition, National Fire Protection Association

ACI 318, 1995 or 1999 Edition, Building Code Requirements for Structural Concrete, American Concrete Institute

ACI 530, 1999 Edition, Building Code for Masonry Structures, American Concrete Institute

ACI 530.1, 1999 Edition, Specifications for Masonry Structures, American Concrete Institute

AISC/ASD Manual of Steel Construction Allowable Stress Design, Ninth Edition, American Institute of Steel Construction, Inc.

AISC/LRFD Manual of Steel Construction Load and Resistance Factor Design, Second Edition, American Institute of Steel Construction, Inc.

NDS, National Design Specification for Wood Construction & National Design Specification Supplement, 1991 or 1997 Edition, ASD, American Forest & Paper Association.

PCI Design Handbook, Fifth Edition, Precast/Prestressed Concrete Institute